Resilient Post Disaster Recovery through Building Back Better

"Building Back Better" (BBB) has been a popular slogan in disaster recovery efforts around the world, including the 2004 Indian Ocean Tsunami, the 2009 Samoan Tsunami, the 2010 Haiti Earthquake and the 2011 Great East Japan Earthquake. BBB has recently been identified as one of four priorities of action for disaster risk reduction globally in the next 15 years by the United Nations *Sendai Framework*. However, there has consistently been a mismatch and confusion in the interpretation of the phrase and what BBB encapsulates which has made proper implementation difficult and unsuccessful at times.

This book explains the concept of "Building Back Better" as an innovative holistic approach to rebuilding a community following a disaster event in order to develop resilience. It begins by exploring the background, development and definitions of BBB. The theory behind establishing BBB as a holistic concept is explained and the internationally recognised BBB Framework developed by the authors is introduced and described. Each of the components of the Framework is explained in detail with findings from international research and case studies from the US, Haiti, Indonesia, Samoa, Sri Lanka, Vanuatu, Gaza, China, Australia, the UK and New Zealand, providing practical recommendations for implementation in recovery projects. There is a focus on the translation of BBB theory into practice to assist implementers to use the BBB Framework and BBB Indicators introduced in this book as effective tools to plan and implement disaster recovery projects.

This publication can be used as a handbook by government, non-governmental and private industry practitioners to prepare for and implement post disaster recovery projects that benefit and strengthen local communities and as a core text on international Disaster and Energy Management courses.

Sandeeka Mannakkara is a post-doctoral research fellow at the University of Auckland's Centre for Disaster Resilience, Recovery and Reconstruction.

Suzanne Wilkinson is Director of the University of Auckland's Centre for Disaster Resilience, Recovery and Reconstruction.

Regan Potangaroa is a Professor of Architectural Science in the School of Architecture at Victoria University, Wellington.

Resilient Post Disaster Recovery through Building Back Better

Sandeeka Mannakkara, Suzanne Wilkinson and Regan Potangaroa

LONDON AND NEW YORK

First published 2019
by Routledge
2 Park Square, Milton Park, Abingdon, Oxon OX14 4RN

and by Routledge
711 Third Avenue, New York, NY 10017

Routledge is an imprint of the Taylor & Francis Group, an informa business

© 2019 Sandeeka Mannakkara, Suzanne Wilkinson and Regan Potangaroa

The right of Sandeeka Mannakkara, Suzanne Wilkinson and Regan Potangaroa to be identified as authors of this work has been asserted by them in accordance with sections 77 and 78 of the Copyright, Designs and Patents Act 1988.

All rights reserved. No part of this book may be reprinted or reproduced or utilised in any form or by any electronic, mechanical, or other means, now known or hereafter invented, including photocopying and recording, or in any information storage or retrieval system, without permission in writing from the publishers.

Trademark notice: Product or corporate names may be trademarks or registered trademarks, and are used only for identification and explanation without intent to infringe.

British Library Cataloguing in Publication Data
A catalogue record for this book is available from the British Library

Library of Congress Cataloging-in-Publication Data
Names: Mannakkara, Sandeeka, author. | Wilkinson, Suzanne
   (Suzanne Jane), author. | Potangaroa, Regan, author.
Title: Resilient post disaster recovery through building back better /
   Sandeeka Mannakkara, Suzanne Wilkinson, Regan Potangaroa.
Description: Abingdon, Oxon ; New York, NY : Routledge, 2019. |
   Includes bibliographical references.
Identifiers: LCCN 2018029652| ISBN 9781138297524 (hbk :
   alk. paper) | ISBN 9781138297531 (pbk : alk. paper) |
   ISBN 9781315099194 (ebk)
Subjects: LCSH: Disaster relief. | Emergency management.
Classification: LCC HV553 .M2667 2019 | DDC 363.34/8—dc23
LC record available at https://lccn.loc.gov/2018029652

ISBN: 978-1-138-29752-4 (hbk)
ISBN: 978-1-138-29753-1 (pbk)
ISBN: 978-1-315-09919-4 (ebk)

Typeset in Goudy
by Swales & Willis Ltd, Exeter, Devon, UK

 Printed in the United Kingdom by Henry Ling Limited

Contents

Acknowledgements · viii

1 Post disaster recovery and the need to Build Back Better · 1

Introduction 1
Disaster recovery and the need to Build Back Better 2
The historical context of disaster management and
   Build Back Better 5
Structure of the book 8

2 Build Back Better theory · 11

Introduction 11
The evolution of Build Back Better 11
Guidelines for Building Back Better 12
Key categories for Building Back Better 14
A Build Back Better Framework 16
Conclusion 26

3 Structural Resilience · 30

Introduction 30
Structural damage and the need for Building Back Better 31
Illustrations of Building Back Better: successes and
   opportunities 32
Improving structural resilience for Building Back Better 39
BBB Indicators for Structural Resilience 40

4 Multi-hazard-based Land-use Planning · 44

Introduction 44
Post disaster land-use planning 45
Illustrations of Building Back Better successes and opportunities 46

Land-use planning for Building Back Better 49
BBB Indicators for Multi-hazard-based Land-use Planning 50

5 Early Warning and Disaster Risk Reduction Education 53

Introduction 53
Illustrations of Building Back Better: successes and opportunities 56
Early Warning and DRR Education for Building Back Better 59
BBB Indicators for Early Warning and DRR Education 59

6 Psychological and Social Recovery 61

Introduction 61
Common post disaster social issues and the need to BBB 62
Illustrations of Building Back Better: successes and opportunities 63
Psychological and Social Recovery for Building Back Better 69
BBB Indicators for Psychological and Social Recovery 70

7 Economic Recovery 73

Introduction 73
Common issues with post disaster economic recovery 74
Illustrations of Building Back Better: successes and opportunities 75
Economic Recovery for Building Back Better 79
BBB Indicators for Economic Recovery 80

8 Institutional Mechanism 84

Introduction 84
Post disaster institutional mechanisms for planning and implementing recovery 85
Illustrations of Building Back Better: successes and opportunities 88
Institutional Mechanism for Building Back Better 97
BBB Indicators for Institutional Mechanism 99

9 Legislation and Regulation 103

Introduction 103
Post disaster legislation challenges 104
Illustrations of Building Back Better: successes and opportunities 105
Legislation and Regulation for Building Back Better 109
BBB Indicators for Legislation and Regulation 111

10 Monitoring and Evaluation 113

Introduction 113
Post disaster monitoring and evaluation 114
Illustrations of Building Back Better: successes and opportunities 115
Monitoring and Evaluation for Building Back Better 116
BBB Indicators for Monitoring and Evaluation 117

11 Building Back Better: from theory to practice 120

Introduction 120
The theory behind Building Back Better 120
Building Back Better in practice 123

Index 128

Acknowledgements

This book is the fruit of a collaborative process. The three authors have been working in multi-disciplinary teams for many years, and as a result have published a substantive body of work on Build Back Better, often working on case studies with different teams. We would like to acknowledge everyone who helped contribute to the research, case studies and contents in this book. In particular, we would like to thank Tinu Rose Francis for advancing the development of the Build Back Better (BBB) Indicators and her extensive work in assessing the recovery of Christchurch following the Canterbury Earthquakes. We thank April Aryal for his BBB research on social capital in Nepal, Maruia Willie and Robert Heather for their BBB research on business recovery in the Cook Islands, Isaac Kim, Ricky Kang, Emma Jarvis and Courtney Buckman for their research on business recovery in Christchurch, Harold Aquino and Mohamed Elkharboutly for their BBB research in Fiji, and all the research teams from the University of Auckland and Victoria University of Wellington who undertook components of research relating to BBB.

We would like to acknowledge the Natural Hazards Research Fund and Ministry of Business, Innovation and Employment for research funding to undertake the Christchurch case study reported throughout this book. We would also like to acknowledge the Centre for Disaster Resilience, Recovery and Reconstruction at the University of Auckland for their support and the many ways they are continuing to make disaster recovery better.

We would like to acknowledge the World Bank's Global Facility for Disaster Reduction and Recovery's extensive BBB country case study assessments which were analysed to extract BBB lessons for this book. We also acknowledge the Agricultural Development Association in Palestine (PARC) and Diakonie Katastrophenhilfe (DKH) reports on Gaza which were included in this book, and thank Mr Rami AbuShaaban (PARC) and Mr Ahmad Safi (DKH) for their critical assistance with the Gaza case study. We acknowledge the Earthquake Engineering Research Institute and World Bank Project team on Build Back Better in Christchurch, parts of which informed the Christchurch sections of this book. We also thank Dr Irshad Timazi and Mr Vickram Chhetri of UNESCO (Pakistan); the Director General of Fiji Red Cross Society, Mr Filipe Nainoca, and the Fiji Red Cross team in Rakiraki (Fiji) including Anaisi Liwa, Meli

Nanuku, Naomi Takatiko, Akesa Buivono, Taitusi Tabaleka, Petaia Saloma, Venina Waiwalu and Seru Sevutia; Habitat for Humanity Fiji, including Mr Masi Latianara, Mr Michael Hill and Mr Kanito Matagasa; Fiji National University's College of Engineering, Science and Technology, led by Mr Salabogi Mavoa, the acting dean; Mr Max Ginn and officials from the Victorian Bushfire Reconstruction and Recovery Authority and Dr Jessica Freame and officials from the Fire Recovery Unit (Australia); Professor Sam Hettiarachchi from the University of Moratuwa and officials from the Disaster Management Centre (Sri Lanka); and all agencies and organisations that assisted us with our country case studies. We also give our thanks to all case study participants who shared their knowledge and experiences to enrich our findings.

Finally, we acknowledge all the studies, reports and documents from which we sourced information. If we have failed to mention anyone else, please let us know so that we may include these in future editions.

1 Post disaster recovery and the need to Build Back Better

Introduction

Despite the increasing number of disasters the world is experiencing, post disaster recovery remains inefficient and poorly managed. This book argues for a systematic approach to disaster recovery which focuses on building resilience through using Build Back Better Principles and embedding these Principles into disaster recovery practices. Post disaster recovery is a time of complex and dynamic decision-making with multiple stakeholders. In this complex environment, the need for clear pathways to create a resilient future is more apparent. Build Back Better provides a recovery process that can deliver resilient solutions to disaster-affected communities. The motivation behind adopting Build Back Better processes is to make communities stronger and more resilient following a disaster event.

Due to the ongoing impacts of disasters on communities, a view emerged where post disaster recovery was seen as an opportunity to not only reconstruct what was damaged and return the community to its pre-disaster state, but to create an opportunity to improve conditions to create a more resilient future (Boano, 2009). Therein the concept of Building Back Better emerged, suggesting that successful recovery of communities following disasters needed to combine rehabilitation and enhancement of the built environment with psychological, social and economic recovery in a holistic way, thus improving overall resilience. The phrase "Building Back Better" became popular during the large-scale reconstruction effort following the Indian Ocean Tsunami disaster in 2004. Building Back Better was officially sanctioned by President Clinton in 2006 with the creation of a set of guidelines to steer recovery and reconstruction activities towards achieving an overall better recovery. More recently, the need to improve recovery with a focus on Building Back Better was given a United Nations mandate under the *Sendai Framework* signed in 2015.

This book starts with the introduction of Build Back Better theory, proposing a framework under which the key features of Building Back Better can be visualised, understood and practically applied. The initial chapters provide an introduction to Build Back Better and show the evolution of the concept into a Build Back Better Framework. Subsequent chapters drill down into individual features, explaining what it means to Build Back Better by providing case studies to illustrate how and when Building Back Better occurs. The book ends with a suite of sophisticated Indicators for Building Back Better to be used by governments, non-governmental organisations (NGOs) and disaster stakeholders for pre-event planning and post-event recovery following disasters.

Disaster recovery and the need to Build Back Better

Disaster management is commonly represented by four phases: mitigation or reduction, preparedness or readiness, response or emergency, and recovery. Governments use disaster management to pre-plan for disaster events, and plan for recovery after disaster events. Build Back Better fits within the pre-event and post-event planning for recovery.

Restoration of the damaged physical, social, economic and environmental impacts of disasters is a complicated and drawn-out process. Traditionally, post disaster reconstruction consisted of simply repairing the physical damage that had been induced by a disaster, and often focused on quick restoration of affected communities. However, authors such as Kennedy et al. (2008) and Lyons (2009) pointed out that rebuilding the built environment and infrastructure exactly as they were prior to a disaster often re-created the same vulnerabilities that existed earlier. If restored to pre-disaster standards, disaster-affected communities would face the same difficulties if exposed to another disaster event in the future. Mitchell (1999), Lewis (2003) and Kijewski-Correa and Taflanidis (2012) noted that the reconstruction and recovery period following a disaster offers an opportunity to address and rectify vulnerability issues found in communities.

Complete recovery requires attention to many different elements. Essentially, the aim of recovery is for governments to work with communities to rebuild their social fabric and their economic status, and reconstruct the natural and built environments. Governments plan for disaster recovery and develop policies that incorporate increased resilience into recovery frameworks. Response and recovery were traditionally seen as a continuum of the same practice. However, increasingly, recovery has been given heightened focus due to large-scale events where recovery has been seen to be problematic. Haas et al. (1977) originally broke down the recovery phase into a Restoration Period of several weeks to a few months, where major services such as communication and transportation are restored; a Replacement Reconstruction Period, where the built environment is restored to pre-disaster levels and social and economic activities are returned to pre-disaster levels or higher; and a Commemorative, Betterment and Developmental Reconstruction Period where memorials and commemoration take place, as well as major construction activities to improve the city for future growth and development. Haas et al. (1977) appear to be the first researchers to include betterment in the recovery process. Wilkinson et al. (2014) produced five stages that disaster environments travel through, showing where the main reconstruction activities occur. The stages proposed by Wilkinson et al. were Chaos, Realisation, Mobilisation, Struggle and the New Normal. Wilkinson et al. explained that after the response stage – where, for instance, the injured and deceased have been taken care of, buildings cordoned off or made temporarily safe, and basic life safety and temporary shelter and food have been attended to – real recovery begins.

However, in the initial stage of recovery there tends to be significant chaos where the general situation is characterised by the question, "What do we do?" The key features found in this stage are the requirement for assessments of needs, identifying priorities and planning for recovery. In the Chaos Stage, initial assessments of damage and the impact of the damage on the community take place. Information-gathering, collation and dissemination, multi-stakeholder interactions and planning arrangements start to be formed. Processes are not always well established, so a certain amount of chaos is evident, causing reassessment and confusion. On the whole, the Chaos Stage often brings into focus the need to undertake a significant rethink of land-use policies, involving questions such as "Should we rebuild?" and "Where?" It is during the Chaos Stage, at the initial planning discussions, that the question of how to improve the resilience of the community for future events and discussions of how to Build Back Better are first addressed.

Following on from the Chaos Stage is the Realisation Stage, which is characterised by the common thought that "the disaster's impact is bigger than we thought". In the Realisation Stage, the common elements that occur are establishing agencies, planning, land-use and re-zoning, new legislation and the introduction of (often quickly produced) new building codes. It is at the Realisation Stage that the desire to Build Back Better, or not Build Back Better, starts to become a reality. Towards the end of the Realisation Stage, newly established recovery organisations emerge, often with planning, financial and welfare functions, all of which generate opportunities to Build Back Better. For the community at large, there is often ongoing anxiety, including temporary accommodation, displacement and minimal or no rebuilding occurring. The sooner good organisational structures staffed with qualified professionals are in place, the better for overall recovery. Davis (2006) believed that:

> many [disasters] concern the relentless pressure for rapid recovery from all quarters which is set against the normal demands for prudent planning, detailed consultation, reviews of safety requirements etc. There is also the demand for reform to be balanced with another pressure for realism or a return to pre-disaster norms.

The Realisation Stage of recovery creates discussion of what to rebuild, where, how and when. While the Realisation Stage is often psychologically difficult, it also provides the opportunity to embed Build Back Better principles into the recovery. Discussions of what to rebuild, where, how and when can be fundamentally reviewed with a Build Back Better perspective: Can we rebuild better than before? Where is the best place to Build Back Better? How can we Build Back Better? This book answers these questions.

Marking the start of the Mobilisation Stage is often the one-year anniversary of the disaster, and the increasing desire to get on with recovery and reconstruction.

The Mobilisation Stage features high energy and increased decision-making, and is a time when Building Back Better starts to become implemented practically. Often public building repairs become evident, as the desire to make physical statements about rebuilding progress are made. During the Mobilisation Stage, the first new buildings emerge, there is high activity in repairs, and also an elevated concern about the wider resourcing problems encountered, leading to increasing costs and shortages of supply. If a holistic approach to Building Back Better is adopted, then the needs of the community, the risks and recovery implementation can create a positive outcome. It is at the Mobilisation Stage that the focus on Building Back Better needs to be maintained. However, there is much less uncertainty from the community in this stage, as structures for reconstruction are in place and increasing activity leads to optimism.

The increased optimism of the Mobilisation Stage is often followed by a Struggle Stage, as the feeling that recovery is hard and not going to plan replaces the optimistic feelings. During the Struggle Stage, what appears is the realisation that there will be no fast recovery. Delays and failures in recovery planning and other inefficient support systems ultimately result in unfulfilled hopes. Recovery takes a very long time, and is always beset with unanticipated problems which slow the recovery timeline. The Struggle Stage sees much activity, with houses and public and commercial properties rebuilt and businesses returning, new business models emerging, but reduced business opportunities. The Struggle Stage sees costs escalating, hardships in procuring materials, skilled professionals in high demand but short supply, and housing affordability problems flowing on from the Mobilisation Stage. The Struggle Stage is very hard for the community, and more people start to reassess options, and there are population shifts from the region. In the pursuit of reconstruction objectives, it is usual for conflicts and discord to ensue between affected groups, government and recovery providers. Such a situation affects morale, and people become susceptible to depression, despondency and emotional exhaustion. Building Back Better addresses the problems occurring in the Struggle Stage by providing evidence for the need for economic stimulation and social support structures to support rebuilding.

The New Normal Stage, which starts many years after the disaster, is characterised by the feeling that "This is how it is, there's no going back." There is always a New Normal Stage following a disaster. Evaluation of this stage can determine whether the environment has created resilience and actually Built Back Better. Disaster recovery and reconstruction do not recreate the same environments seen prior to the disaster. Most environments will be different from those pre-event, including where people live, the way people conduct business, and the types of buildings, architecture and urban form. The New Normal Stage is a time when the community starts accepting the new normal environment. If Building Back Better has created a more resilient community, there should be a sense of security that the buildings and community are safer and future-proofed. A New Normal Stage needs to create communities that are more prepared for future disasters, and with buildings better able to cope. Future-proofing communities is underpinned by the need to use recovery as an opportunity to Build Back Better.

The historical context of disaster management and Build Back Better

Build Back Better, having emerged as a slogan during the 2004 Indian Ocean Tsunami recovery, was defined by Khasalamwa (2009) as a way to utilise the reconstruction process to improve a community's physical, social, environmental and economic conditions to create a more resilient community; where resilience is defined as "the capacity to recover or 'bounce back' after an event" (Twigg, 2007). Build Back Better proposes a broad holistic approach to post disaster reconstruction in order to ensure that affected communities are regenerated in a resilient manner for the future. The underpinning ideas of Build Back Better have been part of international approaches to disaster recovery. Although not specifically advocating for Building Back Better, the *Hyogo Framework for Action* (HFA), which was the outcome of the *Second World Conference on Disaster Reduction* in 2005, emphasised a resilience-based approach to disaster risk reduction and disaster recovery. The HFA was a strategy to make the world safer from natural disasters, focusing on disaster governance, risk identification, assessment, monitoring and early warning, knowledge management and education, reduction of underlying risk factors, and preparedness for effective response and recovery (UNISDR [United Nations Inter-Agency Secretariat for the International Strategy for Disaster Reduction], 2005). Disaster management efforts increased globally after the HFA, which helped nations to understand disaster consequences (Matsuoka et al., 2012). A focus on capacity-building and technology transfer, a multi-hazard approach, taking gender perspectives and cultural diversity into account, and community and volunteer participation were some of the cross-cutting efforts promoted in the HFA.

The first priority of the HFA was governance in disaster management systems. This ensured the development of legislative and institutional frameworks as well as disaster policies for risk reduction to support the HFA priorities (Manyena et al., 2013). The HFA's second priority aimed at reducing risk by identifying, assessing and monitoring disaster risk and establishing early warning systems, promoting a culture of resilience by reducing vulnerability. Promoting knowledge and education was the third priority action, aimed at building a culture of safety and resilience recognising that the impacts of a disaster can be reduced by disseminating hazard information. Risk management and vulnerability reduction was the fourth priority action, which encouraged risk reduction efforts associated with underlying factors such as socio-economic conditions, geography and climate-related factors, and water and land-use conditions. Finally, disaster preparedness and response was the fifth priority action, which ensured strengthening of response mechanisms, encouraging communities to increase their local capacity and equip themselves with knowledge.

6 *Post disaster recovery and the need to BBB*

The post-HFA approach to disaster risk reduction was conceived and drafted at the *Third World Conference on Disaster Reduction* held in Sendai, Japan, in 2015. As a result of thorough assessments and reviews of the HFA and taking account of past disaster lessons, world nations joined together to create an advanced 15-year framework for disaster risk reduction titled the *Sendai Framework for Disaster Risk Reduction* (SFDRR). The SFDRR aims for "the substantial reduction of disaster risk and losses in lives, livelihoods and health in the economic, physical, social, cultural and environmental assets of persons, businesses, communities and countries" over the next 15 years (UNISDR, 2015).

The SFDRR has four priority areas (Figure 1.1). The first focal area is to understand disaster risk in all its dimensions through promoting data collection, ensuring its dissemination, and encouraging the use of this obtained knowledge for pre-disaster assessment, prevention and mitigation, and preparedness and response. The second focus area is to strengthen disaster risk governance to manage disaster risk through dynamic teamwork with a clear vision, plan, competence, coordination and leadership across all stakeholders. It aims to mainstream and integrate disaster risk reduction through policies, laws and regulations and fostering collaboration and partnership across the world to strengthen disaster risk governance. The third priority area promotes public and private investments in disaster prevention and reduction through long-term mitigation measures. It also involves allocation of necessary resources, such as finance and logistics, for the development and implementation of risk reduction strategies, policies, laws and regulations in all relevant sectors. The fourth focus area aims to enhance disaster preparedness for effective response and to "Build Back Better" in recovery, rehabilitation and reconstruction through strengthening disaster preparedness, taking proactive measures for anticipated events and ensuring better response and recovery at all levels.

The SFDRR also focuses on the various stakeholders in disaster management activities. The SFDRR stresses the need to empower stakeholders through education and engagement with disaster risk reduction strategies and plans, which enhances synergies between groups of stakeholders. To implement the SFDRR priorities, it is necessary to provide adequate support to developing nations, promote the use of global technology pools, and ensure access to disaster risk reduction information. Incorporating risk reduction programmes in multilateral and bilateral development assistance programmes across all sectors is also a good means to ensure the achievement of SFDRR goals.

Disaster management ranges from reactive steps to proactive steps, single-agency to partnership, a science-driven to a multidisciplinary approach, response and recovery management to disaster risk management, and planning *for* communities to planning *with* communities (Pearce, 2003). The *Hyogo Framework*, developed in 2005, incorporated disaster prevention, mitigation, preparedness, response and recovery. Ten years later, the *Sendai Framework* further developed recovery thinking, specifically focusing on Building Back Better, describing a model reconstruction and recovery plan for achieving better resilience.

Chart of the Sendai Framework for Disaster Risk Reduction 2015–2030

Scope and purpose

The present framework will apply to the risk of small-scale and large-scale, frequent and infrequent, sudden and slow-onset disasters, caused by natural or manmade hazards as well as related environmental, technological and biological hazards and risks. It aims to guide the multi-hazard management of disaster risk in development at all levels as well as within and across all sectors.

Expected outcome

The substantial reduction of disaster risk and losses in lives, livelihoods and health and in the economic, physical, social, cultural and environmental assets of persons, businesses, communities and countries

Goal

Prevent new and reduce existing disaster risk through the implementation of integrate and inclusive economic, structural, legal, social health, cultural, educational, environmental, technological, political and institutional measures that prevent and reduce hazard exposure and vulnerability to disaster, increase preparedness for response and recovery, and thus strengthen resilience

Targets

| Substantially reduce global disaster mortality by 2030, aiming to lower the average per 100,000 global mortality between 2020–2030 compared to 2005–2015 | Substantially reduce the number of affected people globally by 2030, aiming to lower the average global figure per 100,000 between 2020–2030 compared to 2005–2015 | Reduce direct disaster economic loss in relation to global gross domestic product (GDP) by 2030 | Substantially reduce disaster damage to critical infrastructure and disruption of basic services, among them health and educational facilities, including through developing their resilience by 2030 | Substantially increase the number of countries with national and local disaster risk reduction strategies by 2020 | Substantially enhance international cooperation to developing countries through adequate and sustainable support to complement their national actions for implementation of this framework by 2030 | Substantially increase the availability of and access to multi-hazard early warning systems and disaster risk information and assessments to people by 2030 |

Priorities for Action

There is a need for focused action within and across sectors by States at local, national, regional and global levels in the following four priority areas.

Priority 1	Priority 2	Priority 3	Priority 4
Understanding disaster risk	Strengthening disaster risk governance to manage disaster risk	Investing in disaster risk reduction for resilience	Enhancing disaster preparedness for effective response, and to «Build Back Better» in recovery, rehabilitation and reconstruction

Figure 1.1 Chart of the *Sendai Framework for Disaster Risk Reduction*

Source: Adapted from UNISDR (2015).

8 *Post disaster recovery and the need to BBB*

Structure of the book

The book starts with an analysis of Build Back Better Theory in Chapter 2, which explores the concept of Build Back Better, its origins, background, development and definitions. It stresses Build Back Better as a people-centred concept which strives to create better outcomes for disaster-affected communities and achieve community resilience. The Build Back Better Framework developed by Mannakkara and Wilkinson, which is fast gaining international recognition, is also introduced in this chapter. The key components identified by the Framework: Disaster Risk Reduction (DRR), Community Recovery and Effective Implementation, and their significance for Build Back Better are explained. The subsequent chapters of the book explain the key features of the Build Back Better Framework, with Indicators for each feature.

Chapter 3 describes the development of structural resilience in a community's built environment as part of Disaster Risk Reduction for Building Back Better. The chapter explores the adoption of building codes and regulation, and the often-ignored issues of cost and time-related factors and quality versus speed in the context of Building Back Better. International case studies show the ways in which Building Back Better has assisted disaster recovery. Practical recommendations for implementation of structural resilience demonstrate the Indicators in action.

Chapter 4 shows how multi-hazard-based land-use planning is to be adopted as part of the Build Back Better Framework. The chapter explores how land-use decisions and resettlement can be better aligned with Building Back Better in the best interests of the affected people. Build Back Better Indicators for effective land-use planning are presented.

Chapter 5 gives a description of how early warning and DRR education support Building Back Better by involving the community in DRR and resilience-building.

Chapter 6 examines the importance of supporting the psychological and social recovery of people for overall disaster recovery in the context of achieving Build Back Better. Mechanisms for supporting and empowering affected people and their positive impacts on recovery and resilience are illustrated through case studies.

Chapter 7 explores the economic dimension of Build Back Better. It shows how economic recovery can be best supported in locally affected communities to achieve community resilience. The development of a resilient economic recovery strategy which considers effective funding mechanisms, support for local businesses and protection and enhancement of livelihoods is explored.

Chapter 8 examines appropriate governance systems, and discusses governance for effective implementation of recovery for Building Back Better. Institutional mechanisms, multi-stakeholder management, partnerships, grass-roots-level involvement and quality assurance are covered.

Chapter 9 examines the importance of legislation and regulation for Building Back Better. It illustrates how legislation and regulation can support Building Back

Better through ensuring compliance with Build Back Better-based concepts and improve recovery efficiency through facilitated and fast-tracked procedures.

Chapter 10 describes the importance of monitoring and evaluation throughout the recovery process to identify issues and respond as recovery proceeds, and to extract lessons to improve future disaster management practices. Build Back Better Indicators for Monitoring and Evaluation which serve as practical recommendations for implementation developed through international case study research and literature review are presented, along with case study examples demonstrating these Indicators in action.

Finally, Chapter 11 explores how the Build Back Better Framework and Indicators developed through research can be translated effectively into practice, and presents conclusions.

References

Boano, C. (2009). Housing Anxiety and Multiple Geographies in Post-tsunami Sri Lanka. *Disasters*, 33, 762–785.

Davis, I. (2006). *Learning from Disaster Recovery: Guidance for Decision Makers*. Kobe, Japan: International Recovery Platform. Available: https://www.unisdr.org/files/726_10110.pdf [Accessed 9 August 2018].

Haas, J. E., Kates, R. W. & Bowden, M. J. (1977). *Reconstruction Following Disaster*. Cambridge, MA: MIT Press.

Kennedy, J., Ashmore, J., Babister, E. & Kelman, I. (2008). The Meaning of "Build Back Better": Evidence from Post-tsunami Aceh and Sri Lanka. *Journal of Contingencies & Crisis Management*, 16, 24–36.

Khasalamwa, S. (2009). Is "Build Back Better" a Response to Vulnerability? Analysis of the Post-tsunami Humanitarian Interventions in Sri Lanka. *Norwegian Journal of Geography*, 63, 73–88.

Kijewski-Correa, T. & Taflanidis, A. (2012). The Haitian Housing Dilemma: Can Sustainability and Hazard-resilience Be Achieved? *Bulletin of Earthquake Engineering*, 10, 765–771.

Lewis, J. (2003). Housing Construction in Earthquake-prone Places: Perspectives, Priorities and Projections for Development. *Australian Journal of Emergency Management*, 18, 35–44.

Lyons, M. (2009). Building Back Better: The Large-scale Impact of Small-scale Approaches to Reconstruction. *World Development*, 37, 385–398.

Manyena, S. B., Mavhura, E., Muzenda, C. & Mabaso, E. (2013). Disaster Risk Reduction Legislations: Is There a Move from Events to Processes? *Global Environmental Change*, 23, 1786–1794.

Matsuoka, Y., Joerin, J., Shaw, R. & Takeuchi, Y. (2012). Partnership between City Government and Community-based Disaster Prevention Organizations in Kobe, Japan. In: *Community-based Disaster Risk Reduction*. Bingley, UK: Emerald Group Publishing.

Mitchell, J. K. (ed.) (1999). *Crucibles of Hazard: Mega-cities and Disasters in Transition*. Tokyo, Japan: United Nations University.

Pearce, L. (2003). Disaster Management and Community Planning, and Public Participation: How to Achieve Sustainable Hazard Mitigation. *Natural Hazards*, 28, 211–228.

Twigg, J. (2007). *Characteristics of a Disaster-resilient Community – a Guidance Note*. London: DFID Disaster Risk Reduction Interagency Coordination Group.

UNISDR (2005). *Hyogo Framework for Action 2005–2015: Building the Resilience of Nations and Communities to Disasters*. Kobe, Japan: United Nations Office for Disaster Risk Reduction.

UNISDR (2015). *Sendai Framework for Disaster Risk Reduction 2015–2030*. Sendai, Japan: United Nations Office for Disaster Risk Reduction.

Wilkinson, S., Chang-Richards, A. & Rotimi, J. O. B. (2014). Reconstruction Following Earthquake Disasters. In: M. Beer, E. Patelli, I. Kougioumtzoglou & I. S.-K. Au (eds), *Encyclopedia of Earthquake Engineering*. New York: Springer.

2 Build Back Better theory

Introduction

This chapter develops the theory behind Building Back Better by exploring Build Back Better's origins, background, development and definitions. Build Back Better is a community-centred concept which strives to create better outcomes for disaster-affected communities and achieve community resilience. Existing guidelines that include recommendations for Building Back Better and identify key concepts critical to Build Back Better are analysed. Understanding the dimensions of Building Back Better leads to the Build Back Better Framework developed by Mannakkara and Wilkinson. The Mannakkara and Wilkinson Framework is fast gaining international recognition as a simple way of conceptualising Building Back Better and the multi-factor requirements needed in order to build resilient communities post event. The key Build Back Better components which emerge from the research are identified as Disaster Risk Reduction, Community Recovery and Effective Implementation, and their significance for Building Back Better are explained. Subsequent chapters show the detailed development of Indicators in order to Build Back Better.

The evolution of Build Back Better

The *Sendai Framework for Disaster Risk Reduction* (SFDRR) acknowledged the importance of Build Back Better practices for successful recovery of communities following disasters. Its fourth priority for action specifically aims to enhance disaster preparedness for effective response and to "Build Back Better" in recovery, rehabilitation and reconstruction. Governments under the SFDRR are encouraged to develop pre-event and post-event measures to show how they comply with the SFDRR. Statistics from the United Nations Environment Programme (UNEP, 2008) show an increase in the number of natural disasters over time, attributed to growing populations, urban growth in risk-prone areas due to scarcity of land, and global warming. Along with increasing frequency, recent disasters show an increase in magnitude and resulting destruction (Red Cross, 2010). The SFDRR and the focus on "Building Back Better" create an international mandate for countries in terms of reducing risk, improving recovery and building resilience.

"Build Back Better" first emerged during the multi-national recovery effort following the Indian Ocean Tsunami (Clinton, 2006, Lyons, 2009), specifically as a response to the need to improve current reconstruction and recovery practices and generate safer communities. Against the backdrop of increasing destruction and the need to focus disaster management efforts at the recovery stage, Build Back Better seemed a way of creating hope and reducing vulnerability. The problems of recovery, and the need to focus on Building Back Better, became apparent after various recovery efforts left affected communities in a poorer state than prior to the disaster. Leaving communities in a more vulnerable state was often attributed to the lack of ability to recover well from disaster events. Even in disaster prone areas, pre-existing vulnerabilities were not adequately addressed, leading to acute vulnerabilities in the built environment and exposing communities to greater levels of risk. For instance, the destruction and loss of human lives from the 2005 Kashmir Earthquake in Pakistan was primarily due to the collapse of inappropriately built structures constructed on earthquake-prone land using sub-standard building materials and designed with little earthquake resistance (DN and PA [Duryog Nivaran and Practical Action], 2008, Halvorson and Hamilton, 2010). Poorly planned and ad hoc developments worsened the damage from the Mumbai Floods in 2005. Restoration of the physical, social, economic and environmental damage of disasters is a complicated and drawn-out process. Reconstruction and recovery projects often focus on quick restoration of affected communities, which can replicate and worsen existing vulnerabilities faced by the community (Johnson et al., 2006, Lyons, 2009, TEC [Tsunami Evaluation Commission], 2007). The Tsunami Evaluation Commission *Synthesis Report* (TEC, 2007) provided examples where escalated pressures and the need for fast rebuilding and recovery processes following a disaster further increased the vulnerability of a community. Examples included non-adherence to design and construction policies for buildings and infrastructure; insufficient focus on certain aspects of the recovery process such as livelihood development programmes and small business support programmes; overruling of local government agencies; and neglecting vulnerable groups of people in the community.

The devastation and large-scale reconstruction effort following the Indian Ocean Tsunami in 2004 was the catalyst that gave rise to the concept "Build Back Better".

Guidelines for Building Back Better

Guidelines for Building Back Better started to emerge after the Indian Ocean Tsunami in 2004. Former US President Bill Clinton's (2006) *Key Propositions for Building Back Better* was the first known official document to be published which attempted to provide a comprehensive guideline for implementing Build Back Better ideas in post disaster environments. Clinton (2006) introduced ten propositions for Building Back Better (Figure 2.1).

Build Back Better theory 13

Figure 2.1 Clinton's Build Back Better propositions
Source: Clinton (2006).

The guidelines were similar to principles of recovery that had previously, and have subsequently, been proposed as ways of creating a better community outcome. Many recovery guidelines directly, or indirectly, propose Build Back Better-based recovery and reconstruction operations, whereby the guidelines often advocate for similar approaches, such as risk reduction and community-centric decision-making. For instance, the United Nations Disaster Relief Organization's "Principles for Settlement and Shelter" (Clinton, 2006, Kennedy, 2009) address stakeholder role allocation, needs-based provision of resources to the community, and risk reduction. The Government of Sri Lanka's *Post-tsunami Recovery and Reconstruction Strategy* and "Build Back Better Guiding Principles" (GoSL, 2005b) included needs-based resource allocation and the provision of locally appropriate solutions, community participation and consultation in recovery activities, equity, transparency between stakeholders, risk reduction and consideration of future sustainability, and livelihood support. The US Federal Emergency Management Agency's *Rebuilding for a More Sustainable Future: An Operational Framework* (FEMA, 2000) advocates community-centred recovery operations and hazard-based sustainable risk reduction practices. The Australian Victorian Bushfire Reconstruction and Recovery Authority's "Recovery and Reconstruction Framework" (VBRRA, 2011) focused on the safety and wellbeing of the community, community engagement, equity and tailored solutions. The New Zealand Christchurch Earthquake Recovery Authority's *Recovery Strategy for Greater Christchurch* (CERA, 2016) advocated managing recovery activities using a participatory approach, regenerating the economy, restoring and enhancing the community, reconstruction of the built environment, and restoring natural and healthy ecosystems. Post-Sendai, guidelines for recovery will invariably offer a Build Back Better focus.

Key categories for Building Back Better

The concepts proposed to achieve Build Back Better or better recovery in the various guidelines feature many similarities. Role allocation of stakeholders, community participation and risk reduction appear in most of the guidelines as core concepts or principles that represent the desire to create a better future environment. Developing core concepts creates a focus on how the expected recovery is managed and implemented.

Risk reduction is one of the core concepts that appears in many recovery guidelines, and is a key feature of Sendai. Risk reduction is necessary for Building Back Better. Risk reduction can be seen through the measures put in place to improve the community's physical resilience to natural hazards. Previous post disaster experiences have emphasised the need to identify hazards and determine solutions to reduce risks imposed on people. The Red Cross's *World Disasters Report 2010* (Red Cross, 2010) disclosed that the risks seen in cities are due to a number of reasons, including: growth in informal or illegal settlements; inadequate infrastructure; and building on sites at risk from hazards. The report also stated that many past disasters could have been anticipated and avoided with proper planning, design and construction methods. The Victorian Bushfires Royal Commission *Final Report* (2009 Victorian Bushfires Royal Commission, 2010) recommended the amendment of the Australian Building Code following the Victorian Bushfires, ensuring greater safety standards. The Royal Commission suggested identifying bushfire-prone areas and adopting suitable risk reduction methods, such as improved building and planning controls and improving warning systems. The National Mitigation Strategy produced in Turkey following the Kocaeli and Duzce earthquakes of 1999 also stated the need for site-specific hazard identification before reconstruction, as well as retrofitting and updating structural codes and using tax incentives to encourage mitigation work (Bakir, 2004). A 2008 *South Asia Disaster Report* (DN and PA, 2008) recommended that hazard and vulnerability maps be produced and building codes enforced to avoid development-related disasters in the future. The suggestions show that risk reduction can primarily be achieved through improving the structural designs of the built environment to enhance its ability to resist damage from disasters, through hazard-based land-use to avoid or manage risks, and through education and early warning systems.

Most of the guidelines place the community at the heart of recovery. The focus on the community leads guidelines to suggest areas of focus such as including the community in recovery, empowering the community, providing recovery solutions based on community needs, considering social aspects of recovery, and enhancing and supporting psychological recovery. *Community recovery* deals with community-relevant aspects in recovery. Aspects such as improving the social and economic conditions of the community for long-term sustainability by supporting livelihoods and regenerating the economy, providing needs-based, locally and culturally appropriate recovery solutions and focusing on community wellbeing create better community recovery. Chamlee-Wright and Storr (2009), Chang (2010) and Kennedy (2009) believed that keeping the community together and involving them in collective activities (such as social gatherings and participation in reconstruction and recovery work) and providing psychological support improved overall recovery. The recovery effort following the Australian Victorian Bushfires was a good attempt at psycho-social recovery of affected people through the provision of "case managers" for each family to provide individualised information and resources to support recovery, as well as through providing services such as counselling, youth support, children support, men's getaways, memorial services and community events (VBRRA, 2010).

A core theme which encapsulates the idea of Build Back Better is that the community should drive recovery and that recovery operations require consultation with and participation of the community. Recovery activities are for the benefit of the affected local community. The needs, dynamics, culture and other pre-existing socio-political, environmental and physical issues of the community need to be considered (Khasalamwa, 2009, DN and PA, 2008, James Lee Witt Associates, 2005, Olshansky, 2005). A decentralised approach empowers communities and provides a greater level of satisfaction about the outputs (Davidson et al., 2007, Lyons, 2009).

Clinton (2006) stated that "a sustainable recovery process depends on reviving and expanding private economic activity and employment and securing diverse livelihood opportunities for affected populations". The need for economic rejuvenation and encouraging the community to return to their former livelihoods or venture into new ones has been reinforced by many authors, including Haigh et al. (2009), Bredenoord and van Lindert (2010) and Johnson et al. (2006). Economic recovery and livelihood regeneration can be achieved through initiatives such as cash-for-work programmes (paying locals to become involved in reconstruction work) (Haigh et al., 2009), skills training programmes (James Lee Witt Associates, 2005), owner-building schemes (Bredenoord and van Lindert, 2010, Johnson et al., 2006, Lyons, 2009), providing job opportunities and sustainable livelihood options (Monday, 2002, Twigg, 2007) and arranging financial help and grants for small businesses and micro-enterprise schemes (Asian Development Bank et al., 2005, GoSL, 2005a). Red Cross (2010), Batteate (2006) and Winchester (2000) stated that successful livelihood recovery programmes utilise grass-roots schemes and attempt to understand the requirements of the community to re-establish the local economy and livelihoods. Building communities Back Better requires a range of activities focussed on improving wellbeing and livelihood.

A commonly arising issue in post disaster environments is the difficulty in coordinating with the large number of stakeholders involved and determining their specific roles to avoid duplication of activities (DN and PA, 2008, GoSL and UN, 2005, James Lee Witt Associates, 2005). Clinton's propositions state that government officials should pre-plan for disasters by considering ways to organise government agencies and institutions, with clarification provided on roles and responsibilities as well as forming partnerships with other organisations. The creation of overseeing bodies to coordinate stakeholders is a common feature of recovery environments. In order to operate effectively, legislative and regulative measures are required to facilitate reconstruction and recovery and reduce future risk. For instance, legislation can be used to prohibit and/or control construction in hazard-prone areas (Bakir, 2004). Haigh et al. (2009) identified the need for legislation to be simplified and streamlined to assist recovery operations and reduce delays. The creation of national policies for employment creation (Boano, 2009) and resettlement (Frerks and Klem, 2005, GoSL, 2005a) have also been proposed based on previous disaster experiences. The implementation of post disaster reconstruction and recovery activities to Build Back Better therefore requires a focus on who will be involved in decision-making, including identification of stakeholders and their roles, relationships and responsibilities, and legislative and regulatory measures to reduce risks, facilitate recovery and provide policies for social and economic recovery.

The need to take lessons learnt from disaster experiences and translate them into recovery plans and training programmes to prepare for future events is part of Building Back Better. The 2009 Victorian Bushfires Royal Commission (2010), Clinton (2006), Haigh et al. (2009) and Halvorson and Hamilton (2010) all stress the importance of monitoring and evaluating recovery efforts and producing future recovery plans to create a resilient community that has the knowledge and resources to respond to a future disaster event. Monitoring and evaluation exercises need to be undertaken through all recovery activities. A successful recovery requires effective and efficient recovery solutions as part of Building Back Better. In implementing monitoring and evaluation, the recovery can be flexible to changing community needs, and respond accordingly.

A Build Back Better Framework

Core categories enable the creation of a framework that can be used to understand Building Back Better. Figure 2.2 shows the Build Back Better Framework.

Build Back Better theory 17

Figure 2.2 International Build Back Better Framework

Source: © Mannakkara and Wilkinson, University of Auckland, 2010, 2016.

Key categories in the Build Back Better Framework

The concepts proposed to achieve Build Back Better during reconstruction and recovery in the various guidelines in the previous section feature similarities. The key categories introduced in the guidelines for improving reconstruction and recovery efforts and Building Back Better include: Disaster Risk Reduction, Community Recovery and Effective Implementation.

Disaster Risk Reduction

Disaster Risk Reduction identifies all actions taken towards reducing disaster risks in communities to improve the physical resilience in the built environment. Previous post disaster experiences have emphasised the need to identify prevalent hazards and determine solutions to be undertaken to reduce risks imposed on people.

The primary methods of Disaster Risk Reduction are identified in the BBB Framework through three BBB Principles: (1) Structural Resilience through improving structural designs, (2) Multi-hazard-based Land-use Planning for improved land-use planning, and (3) Early Warning and DRR Education to implement effective early warning mechanisms and provide education on risks. The importance of reviewing and changing building designs and codes to improve the structural integrity of buildings and infrastructure following a disaster is widely understood, but is less frequently attained successfully in practice due to a range of common issues. Poor regulative powers and the lack of strict enforcement can lead to building code changes being disregarded, resulting in sub-standard structures in the rebuild. When the Indian Ocean Tsunami struck, enforcement of building codes was mainly restricted to urban and suburban areas in Sri Lanka. The rural and coastal areas were the main victims of the disaster, where the lack of strict structural standards resulted in magnified damage (Pathiraja and Tombesi, 2009). Extra costs incurred by adopting new technologies and materials to improve structural resilience also discourage compliance with new building codes worldwide (Batteate, 2006).

The experiences of post disaster reconstruction efforts have provided lessons which can be adopted when implementing structural changes to avoid the above-mentioned issues. Build Back Better theory suggests that hazard-based building regulations should be created using multi-hazard assessments in areas chosen for redevelopment and reconstruction. Consistent regulations and a strong legal framework are necessary to assist the adoption of building codes and regulations and ensure that structural changes improve the built environment (Clinton, 2006). As structural improvements are expensive and unaffordable, especially in post disaster settings, long-term funding needs to be made available to cover extra costs for structural improvements and promote adoption. Quality of reconstruction can be maintained by arranging inspections during construction. Stakeholders involved in the rebuild, such as builders, engineers and architects, should be trained on revised building codes and other specific requirements to avoid inconsistencies and produce good-quality results in order to Build Back Better.

A land-use planning strategy was used in the post disaster recovery efforts following the Indian Ocean Tsunami and the Samoan Tsunami, resulting in the relocation of coastal communities further inland to avoid future impacts of coastal hazards (Kennedy et al., 2008). The mandatory resettlement operations in Sri Lanka and Samoa were problematic due to the lack of consideration given to the lifestyles of the local people, which led to the loss of their sea-dependent livelihoods, dissatisfaction with their new settlements, and illegal return of people to the original coastal lands (Kennedy et al., 2008). A recurring issue with relocation is the focus on moving communities away from a certain hazard, resulting in exposing communities to new, unanticipated hazards. Well-intended land-use planning measures can also fail due to the lack of knowledge and awareness of local people who do not comply with new regulations, and the lack of experience and knowledge of local governing authorities who do not enforce new regulations (Kennedy et al., 2008).

Therefore, it was recommended by Baradan (2006) that hazard assessments of current land sites and possible new land sites should be carried out, after which risk zone maps are to be created which divide the land into zones based on the level of risk. Appropriate land-uses and new planning and building regulations based on the risk zone maps are to be created. The risk zone maps should be legislated and included in council development plans and approval permit procedures to ensure compliance. Examples such as Taiwan's Mitigation

Plans, the Philippines Municipal Maps and the Christchurch City Plan in New Zealand following the Canterbury Earthquakes display successful application of Build Back Better measures to create safer developments. If resettlement to lower-risk lands is opted for, Mannakkara and Wilkinson (2012b) recommend that a comprehensive resettlement strategy should be created with community consent, which takes into account risk levels of new lands, community preferences and livelihood and lifestyle opportunities offered in the new locations, for resettlement to be a success.

DN and PA (2008) encourage educating communities about risks and the importance of risk reduction measures and engaging them in collective risk reduction efforts. The Participatory Flood Risk Communication Support System (Pafrics) developed in Japan to educate locals and other stakeholders including NGOs and local governments on flood risks and risk management strategies is an example of a participatory tool.

Community Recovery: through psychological and social recovery

The BBB Framework identifies the first approach to Community Recovery through the BBB Principle of Psychological and Social Recovery. Supporting psycho-social recovery of affected communities has been identified as essential for Building Back Better (Davidson et al., 2007). Post disaster recovery often focuses on providing fast solutions in an attempt to re-establish a sense of normality in affected communities as soon as possible (Khasalamwa, 2009). The focus on speed results in overlooking the real needs of communities. Communities are often not consulted to provide their input on reconstruction and recovery (Boano, 2009). The lack of community consultation and participation leads to the provision of recovery solutions that are not suitable. For example, some of the new houses constructed in Sri Lanka by humanitarian agencies during the Indian Ocean Tsunami rebuild featured bathrooms with half-height walls and shared bathrooms for males and females, which were culturally unacceptable (Ruwanpura, 2009). Locals were unhappy with the reconstruction of homes following the 1999 Marmara Earthquake in Turkey as their local life, culture and aesthetics were not considered. Khasalamwa (2009) stated that insufficient attention to social, cultural and ethnic facets of communities during recovery can exacerbate pre-existing vulnerabilities. Separation during disasters and resettlement operations disrupts community cohesion and psychological recovery (Florian, 2007).

Social issues arising in post disaster environments are primarily related to social/cultural/religious/ethnic factors and psychological factors. Reconstruction is a chaotic and stressful time for individuals who are also experiencing trauma. These communities require various forms of assistance as part of Building Back Better. Personalised advice and one-on-one support provided to families in Columbia during the 1999 earthquake recovery were a success. Similar forms of personal assistance were provided during the Victorian Bushfire recovery in Australia. James Lee Witt Associates (2005) recommended arranging specialised assistance for vulnerable communities. Providing psychological support and counselling is essential during recovery. The establishment of information centres which offer easy access to recovery-related information for the community is also recommended. Upholding a sense of community spirit and improving community cohesion through organising group activities is recommended for social recovery. The Canterbury Earthquake recovery in Christchurch proposed sports, recreation, arts and cultural programmes to engage the community and provide a sense of normality.

One of the first steps to be taken in post disaster recovery efforts in order to Build Back Better is to understand the local context of the affected community through needs assessments and surveys in order to provide appropriate assistance to satisfy the community (Khasalamwa, 2009). The reconstruction and recovery policies must then be developed based on local requirements to support and preserve the local culture and heritage. Batteate (2006) stated that maintaining community involvement throughout recovery is integral for Build Back Better success. The importance of decentralisation to empower disaster-affected communities by enabling them to take responsibility for the recovery effort and become involved in decision-making has been stressed in the literature. The establishment of community consultation groups is an effective way to communicate with the community. Community consultation groups consisting of community leaders from pre-existing community groups and reputable members of the community to liaise between the wider community and governmental authorities have been successful in Sri Lanka and India. Existing community groups can also be called upon to assist with recovery activities.

Community Recovery: through Economic Recovery

The second approach to Community Recovery presented in the BBB Framework is the BBB Principle of Economic Recovery. Supporting economic recovery of the community and supporting livelihood regeneration and entrepreneurship is also an important part of recovery. Disasters cause damage to the economy of communities, with the disruption of businesses and income-generating industries leading to issues such as high inflation rates and poverty. The adverse effects of disasters on the economy can also impede the overall recovery of a city. Hurricane Katrina displayed a disaster's long-term impacts on higher education and health care in New Orleans, which were the foundations of the city's economy, eventually leading to a decline in population numbers as people moved away in search of better opportunities.

Post disaster recovery efforts to date have shown support for economic recovery with strategies such as "cash-for-work" programmes, provision of business grants, "asset replacement" programmes to provide industries with necessary resources, and training programmes to up-skill locals and help them find work. In Aceh, Indonesia tsunami-affected people were trained and employed in reconstruction to provide them with a source of income alongside the opportunity to become involved in their own recovery (Kennedy et al., 2008). In Japan following the 2011 earthquake and tsunami, the government decided to consolidate smaller fishing markets into large fishing centres to enable fishermen to support each other (Okuda et al., 2011). The New Zealand Christchurch City Council's Central City Plan proposed fast-tracking of building consents for businesses to allow quicker repair and construction work. Despite the implementation of such initiatives, post disaster economic recovery is reportedly slow and below pre-disaster levels. The lack of success in economic recovery initiatives can be attributed to insufficient backing from policies and legislation for employment creation and lack of sufficient consideration given to the needs of affected communities.

Clinton (2006) said in his Build Back Better propositions that "a sustainable recovery process depends on reviving and expanding private economic activity and employment and securing diverse livelihood opportunities for affected populations". Thus, the uniqueness of Build Back Better comes from the integrated approach it proposes by giving economic recovery as much importance as reconstruction and aiming to provide solutions to suit local dynamics and preferences.

Monday (2002) stated that one of the first steps needed for effective economic recovery is to obtain accurate information about the local population through data collection and consultation with local governmental authorities, and a comprehensive economic recovery strategy must be created that is tailor-made to suit each different community based on data obtained. Where applicable, attractive and flexible low-interest loan packages, business grants and resources should be provided to support the livelihoods of the disaster-affected. Training programmes should be set up to support people in improving their existing livelihoods or acquire new skills. Mannakkara and Wilkinson (2012a) propose that business support and counselling services should be provided to assist with the economic recovery. Rebuilding of businesses must also be facilitated through special fast-tracked permit procedures. Incentives such as subsidised accommodation must be provided to attract builders from other areas to participate in rebuilding.

Effective Implementation

A successful recovery effort requires effective and efficient recovery solutions as part of Building Back Better. The BBB Framework identifies three ways in which the efficiency and effectiveness of post disaster recovery can be improved through three BBB Principles: (1) better management of stakeholders by selecting an appropriate Institutional Mechanism, (2) through the use of appropriate post disaster Legislation and Regulation, and (3) through the use of Monitoring and Evaluation mechanisms.

One of the most common issues with post disaster environments is the difficulty in coordinating between stakeholders to produce a unified outcome. Initially, there is often no organisation in charge of the overall recovery effort. The lack of guidance leads different stakeholders to participate disjointedly, promoting personal agendas which conflict with the interests of the local community (Batteate, 2006). For example, non-governmental organisations which operated in Sri Lanka following the Indian Ocean Tsunami constructed homes that were unsuitable for locals and were largely abandoned. The pressure for fast results during recovery also prevents well-intentioned stakeholders from considering community needs. Ambiguity about the roles of different stakeholders is another issue. The 2009 Victorian Bushfires Royal Commission (2010) stated that the roles of personnel involved in the recovery effort were unclear, which led to the duplication of some activities. Many stakeholders involved in recovery have no previous experience in post disaster environments, leading to ad hoc responses (Kennedy, 2009). Often post disaster interventions are instigated by the national government without sufficient consultation or power given to local councils (Clinton, 2006). Local-level organisations with useful local knowledge lack the capacity to operate to their full extent when impacted by disasters and are therefore excluded from recovery efforts. The lack of proper role allocation, coordination and involvement of local-level stakeholders are common issues found in post disaster reconstruction environments.

A step taken to improve the management of large numbers of stakeholders in major disasters in order to Build Back Better is the creation of a separate body to act as a recovery authority. Examples of recovery authorities created to manage reconstruction include: the Bureau of Rehabilitation and Reconstruction (BRR) in Indonesia following the Indian Ocean Tsunami; Bam's Reconstruction Supreme Supervisory and Policymaking Association (BRSSPA) in Iran following the 2003 Bam Earthquake; the Victorian Bushfire Reconstruction and Recovery Authority (VBRRA) in Australia following the 2009 Victorian Bushfires; and the Canterbury Earthquake Recovery Authority (CERA) following the 2010 and 2011 Canterbury earthquakes in New Zealand. The recovery agencies contributed to the success of recovery to differing extents. Clinton (2006) said that stakeholders must operate with a common set of standards, approaches and goals in order for recovery to be a success. Twigg (2007) proposes that the recovery authority should be responsible for establishing clear roles and responsibilities for the different stakeholders to divide recovery tasks based on resources and skills and avoid duplication.

Functional partnerships and linkages established between organisations can enhance reconstruction projects. Post disaster recovery is a unique environment which requires deviation from normal procedures. Information-sharing between organisations is one such deviation. The Federal Emergency Management Agency in the United States advocates the sharing of information, contacts, resources and technical knowledge between organisations to help recovery activities (FEMA, 2000). Knowledge from past disasters should be retained and transferred to the government and other relevant organisations that will be involved in future post disaster efforts. Twigg (2007) recommends that local government should be included as a key stakeholder in the recovery effort and also be given the responsibility to manage local-level activities.

Another obstacle preventing successful Build Back Better-centred recovery is the absence of proper controls to enforce Build Back Better Principles. Having Build Back Better knowledge and producing recovery plans in line with these Principles is futile without proper legislation and regulations in place to ensure they are implemented. A common challenge in post disaster environments is the sudden increased workload, especially in the building industry, along with a drop in the workforce across local organisations, which slow down and impede recovery activities. Post disaster reconstruction requires time-consuming activities such as hazard analysis, land selection, infrastructure development and rebuilding to be done in a relatively short period of time. Legislation that is customarily used to impose security and safety controls (such as building consents) can become an obstacle in high-pressure post disaster environments. Time-consuming procedures, insufficient resources to process permits and the lack of fast-tracked methods delay reconstruction. Delays in permits was a major reason for the hold-up in housing repair and rebuilding following the 2005 Bay of Plenty storm in New Zealand (Middleton, 2008). It is important to facilitate recovery-related activities by simplifying, fast-tracking and exempting certain rules and regulations using special legislation. Legislation can be used to remove unnecessary red tape to facilitate recovery activities. Meese III et al. (2005) reported a good example in the recovery following the 1994 Northridge Earthquake in the USA, where legislative suspensions and emergency powers greatly reduced highway reconstruction time. The construction work provided employment, and opening up the highways soon after the disaster helped boost the economy.

Post disaster legislation can be used to ensure compliance with Build Back Better-based activities as well as to facilitate normal operations to improve the efficiency of recovery efforts. The lack of enforcement of hazard-related laws and adequate risk-based building controls contributed to the large-scale devastation caused by the 2004 Indian Ocean Tsunami (DNS [Duryog Nivaran Secretariat] and PA, 2005). The same was seen in countries like Pakistan, Turkey, Samoa and Haiti. Enforcing updated risk-based building design standards through the use of compulsory building codes and maintaining construction standards through careful inspections are important regulatory requirements in reconstruction (James Lee Witt Associates, 2005). Lack of awareness and understanding of new legislation can also lead to non-compliance. In the post-tsunami recovery effort in Sri Lanka, external NGOs that took part did not comply with local standards due to lack of awareness (Boano, 2009). The *National Post-tsunami Lessons Learned and Best Practices Workshop* held in Sri Lanka in 2005 highlighted the importance of training stakeholders (especially external NGOs) about existing and newly introduced legislation and regulations (GoSL and UN, 2005). The community's support can also be obtained by educating them about legislation and regulations that must be adhered to in reconstruction and recovery.

The effectiveness and efficiency of post disaster reconstruction and recovery activities is crucial to the success of a community's restoration following the impact of a disaster event. Having knowledge of Build Back Better concepts in designing recovery programmes is insufficient without systems in place to oversee and monitor implementation. The creation of a recovery strategy to assist in conducting post disaster reconstruction and recovery activities is a common response following disaster events. Despite having recovery strategies and revisions in legislation and regulation to improve recovery activities, the findings by Tas (2010) indicated that compliance was not monitored in the respective recovery efforts in Sri Lanka and Turkey, leading to poorly executed recovery projects. The lack of properly trained professionals who were competent in post disaster environments and disaster management activities seriously affected the outcome of recovery efforts. The shortage of effective information- and knowledge-sharing and dissemination are also reasons for unsatisfactory disaster management practices. Findings from the Business Civic Leadership Centre in 2012 on "What a Successful Recovery Looks Like" raised concerns that long-term recovery beyond reconstruction often does not take place due to the lack of mechanisms and expertise, which prevent affected communities from satisfactorily "Building Back Better" in the long run.

Recommendations to improve post disaster recovery efforts through monitoring and evaluation have been provided in much of the literature. The role of monitoring and evaluation is twofold: (1) to monitor and ensure compliance of recovery activities in accordance with the recovery strategy in place and relevant guidelines and regulations (Clinton, 2006) and (2) to obtain lessons for the future and improve future disaster management and post disaster reconstruction and recovery efforts (Monday, 2002).

The 2003 Bam Earthquake reconstruction provided a good example where rebuilding was monitored by providing construction supervision which assisted in assuring the quality of the rebuild. Clinton (2006) stated that the Tsunami Recovery Impact Assessment and Monitoring System (TRIAMS) was put in place during the Indian Ocean Tsunami recovery for the most affected countries. The recovery strategy in Christchurch, New Zealand has also been equipped with monitoring mechanisms. Clinton (2006) suggested that long-term recovery should be monitored through continued data collection to ensure that recovery efforts do not leave communities with residual issues.

Monday (2002) pointed out that monitoring can be used to identify problems with post disaster interventions and establish lessons learnt. Lessons learnt should be incorporated into revising policy and procedures for future disaster management practices. Bakir (2004) recommends that public education campaigns should be run on lessons learnt, including the community in participatory disaster management. Public seminars have been held and advice notes have been distributed in Australia during the Victorian Bushfires recovery to keep the community informed about revised guidelines and standards. Workshops have been held in the Philippines, Japan and California involving the community in vulnerability identification, which have been successful (Batteate, 2006).

Different sets of guidelines have been produced to assist Building Back Better, including: ten propositions by former US President Bill Clinton; principles for settlement and shelter by United Nations Disaster Relief Organization (UNDRO, 1982), guiding principles established locally in Sri Lanka during post-tsunami reconstruction (Disaster Relief Monitoring Unit of the Human Rights Commission of Sri Lanka, 2006), and a variety of reports.

Importance and implications of the Build Back Better Framework

Despite having several Build Back Better guidelines, the implementation of Build Back Better in post disaster practices in recent disasters such as the 2009 Victorian Bushfires and the 2010 Haiti Earthquake has had shortcomings.

Although knowledge of Build Back Better concepts exists and recovery plans are produced for guidance, there are complications in post disaster environments, such as: balancing the extent of improvement made to structural designs and land-use plans for risk reduction with affordability, time constraints, preferences and traditions of the local community (Baradan, 2006, Boano, 2009, Clinton, 2006, Tas, 2010); coordinating and communicating between the numerous stakeholders involved to avoid duplication of activities and produce efficient results (2009 Victorian Bushfires Royal Commission, 2010, Batteate, 2006, Khasalamwa, 2009); establishing programmes to regenerate and re-establish livelihoods of the people to match their skills, resources and future demands (Khasalamwa, 2009, Clinton, 2006, James Lee Witt Associates, 2005); and facilitating reconstruction and recovery activities to take place for speedy results without compromising quality (James Lee Witt Associates, 2005, Khasalamwa, 2009, Ozcevik et al., 2009). The issues listed above have not been properly dealt with in order to overcome them. For example, although risk reduction through improvement of structural designs and hazard-based land-use planning is proposed by many sources, issues regarding the affordability and practicality of adopting the enforced structural improvements; and scarcity of land which may restrict the ability to prevent developments in high-risk lands have not been considered.

The numerous guidelines currently available providing instructions for Build Back Better can be confusing for practitioners, making it difficult to determine which guideline to follow, which has thus inhibited adoption up to now. With the creation of the Build Back Better Framework and the recognition of an inclusive set of Build Back Better Principles, it is possible to determine solutions for previously overlooked but commonly encountered issues in reconstruction and recovery while having an overarching outlook on all the different aspects involved. The Build Back Better Framework and its Build Back Better categories and Principles can be used as a starting point to develop recommendations to improve the applicability of Build Back Better concepts in the areas of risk reduction, community recovery and effective and efficient implementation. Build Back Better may be successfully achieved in post disaster environments through having a prescriptive, straightforward, comprehensive framework along with legal and governmental backing. As a practitioner in either government, non-governmental or private institutions, the Build Back Better Framework and Build Back Better Principles serve as a guide to design post disaster recovery programmes and reconstruction and recovery plans. Individual projects within recovery programmes can be made successful by considering the different areas required for Building Back Better using Build Back Better Principles. The Principles need to be looked at in a holistic manner when designing an overall recovery programme to gain an understanding of the components that need to be included in the recovery effort. It is important to place emphasis on the success of the overall programme as a result of better management and Build Back Better consideration given to individual projects placed under the programme. It is necessary to establish a clear understanding and links between the projects within the programme to ensure overall success. Build Back Better is an important concept which incorporates adopting a holistic approach to improve a community's physical, social, environmental and economic conditions during post disaster reconstruction and recovery activities to create a resilient community.

Conclusion

Build Back Better is an important concept for post disaster reconstruction and recovery, signifying the need to use reconstruction as an opportunity to not only recover from the encountered disaster, but to improve the resilience of communities to face and withstand future disaster events. Build Back Better represents adopting a holistic approach towards recovery by addressing risk reduction of the built environment, psycho-social recovery of affected people, and rejuvenation of the economy in an effective and efficient manner.

The various concepts and suggestions presented in the existing literature to Build Back Better have not made its implementation straightforward. Analysis of key literature on post disaster reconstruction and recovery with suggestions to improve post disaster practices to Build Back Better led to the identification of three key categories and eight principles which depict Build Back Better.

The first category is Disaster Risk Reduction, and it can be achieved primarily through the improvement of structural designs in buildings and infrastructure, better risk-based land-use planning and early warning and DRR education provided to stakeholders and the community. Build Back Better requires improved building codes and land-use plans to be enforced using a strong legal framework along with financial backing to encourage adoption. Quality assurance of the rebuild is also integral for Building Back Better. Communities also need to be equipped with effective early warning systems, have a good awareness of their risks and take part in risk reduction exercises.

The second category for BBB is Community Recovery, which includes psycho-social recovery and economic recovery. Psycho-social recovery needs to be addressed to assist communities with moving forward with their lives as an important part of overall community recovery. Psycho-social recovery of affected people needs to be assisted through the provision of support services such as personal case management, counselling, and social activities. Inclusion of community members in recovery activities is another way to support psycho-social recovery and provide recovery solutions that are in line with community needs as part of Building Back Better.

Economic Recovery is essential for the recovery of communities. An informed economic strategy to address and support community-specific issues is the first step towards Build Back Better-based economic recovery. Financial assistance, training and business rebuilding support need to be provided to assist with economic recovery.

The third category for BBB is Effective Implementation. Effective Implementation refers to having an appropriate institutional mechanism in place for an effective and efficient recovery process, along with necessary legislation and regulation and monitoring and evaluation. The creation of a recovery authority to allocate roles and coordinate and manage stakeholders is recommended. Successful recovery requires local-level partnerships and contributions to provide locally viable recovery solutions. Compliance with Build Back Better-based concepts in recovery needs to be ensured through the use of appropriate post disaster legislation and regulation to enforce risk reduction and community recovery initiatives. Legislation and regulation can also be used to facilitate post disaster recovery activities by fast-tracking and exempting normal procedures.

The effective implementation of risk reduction and community recovery initiatives concurrently will result in Building Back Better. Recovery efforts also need to be monitored continuously through short-term and long-term recovery to ensure compliance with Build Back Better concepts and to obtain lessons to improve future disaster management efforts.

It is intended that the Build Back Better Framework and Build Back Better categories and Principles will be utilised to provide solutions for commonly encountered complications in reconstruction and recovery activities to make Building Back Better possible.

References

2009 Victorian Bushfires Royal Commission (2010). *Final Report – Summary*. Melbourne, Australia: 2009 Victorian Bushfires Royal Commission.

Asian Development Bank, Japan Bank for International Cooperation & World Bank (2005). Preliminary Damage and Needs Assessment. In: *Sri Lanka 2005 Post-Tsunami Recovery Programme*. Colombo, Sri Lanka: Asian Development Bank, Japan Bank for International Cooperation and World Bank.

Bakir, P. G. (2004). Proposal of a National Mitigation Strategy against Earthquakes in Turkey. *Natural Hazards*, 33, 405–425.

Baradan, B. (2006). Analysis of the Post-disaster Reconstruction Process Following the Turkish Earthquakes, 1999. In: Group, I. R. (ed.) *International Conference on Post-Disaster Reconstruction: Meeting Stakeholder Interests*. Montreal, Canada: University of Montreal.

Batteate, C. (2006). Urban Disaster Risk Reduction and Regeneration Planning: An Overview. *Focus: Journal of the City and Regional Planning Department*, 3, 11–17.

Boano, C. (2009). Housing Anxiety and Multiple Geographies in Post-tsunami Sri Lanka. *Disasters*, 33, 762–785.

Bredenoord, J. & van Lindert, P. (2010). Pro-poor Housing Policies: Rethinking the Potential of Assisted Self-help Housing. *Habitat International*, 34, 278–287.

CERA (2016). *Recovery Strategy for Greater Christchurch*. Available: http://ceraarchive.dpmc.govt.nz/documents/recovery-strategy-greater-christchurch [Accessed 9 August 2018].

Chamlee-Wright, E. & Storr, V. H. (2009). Club Goods and Post-disaster Community Return. *Rationality and Society*, 21(4), 429–458. Available: http://rss.sagepub.com/content/21/4/429 [Accessed 9 August 2018].

Chang, K. (2010). Community Cohesion after a Natural Disaster: Insights from a Carlisle Flood. *Disasters*, 34, 289–302.

Clinton, W. J. (2006). *Lessons Learned from Tsunami Recovery: Key Propositions for Building Back Better*. New York: Office of the UN Secretary-General's Special Envoy for Tsunami Recovery.

Davidson, C. H., Johnson, C., Lizarralde, G., Dikmen, N. & Sliwinski, A. (2007). Truths and Myths about Community Participation in Post-disaster Housing Projects. *Habitat International*, 31, 100–115.

Disaster Relief Monitoring Unit of the Human Rights Commission of Sri Lanka (2006). Building Back Better: Way Forward. *National Workshop on Guiding Principles, 2006*. Colombo, Sri Lanka: Practical Action – South Asia Programme.

DN & PA (2008). *Disaster and Development in South Asia: Connects and Disconnects. South Asia Disaster Report*. Colombo, Sri Lanka: Duryog Nivaran and Practical Action.

DNS & PA (2005). *Tackling the Tides and Tremors: South Asia Disaster Report 2005*. Colombo, Sri Lanka: Duryog Nivaran Secretariat and Practical Action – South Asia Programme.

FEMA (2000). *Rebuilding for a More Sustainable Future: An Operational Framework*. Washington, DC: Federal Emergency Management Agency.

Florian, S. (2007). Housing Reconstruction and Rehabilitation in Aceh and Nias, Indonesia – Rebuilding Lives. *Habitat International*, 31, 150–166.

Frerks, G. & Klem, B. (2005). *Tsunami Response in Sri Lanka: Report on a Field Visit from 6–20 February 2005*. Wageningen, The Netherlands: Wageningen University and Clingdael University.

GoSL (2005a). *Post-tsunami Recovery and Reconstruction: Joint Report of the Government of Sri Lanka and Development Partners*. Colombo, Sri Lanka: Government of Sri Lanka.

GoSL (2005b). *Post-tsunami Recovery and Reconstruction Strategy*. Colombo, Sri Lanka: Government of Sri Lanka.

GoSL & UN (2005). *National Post-tsunami Lessons Learned and Best Practices Workshop*. Colombo, Sri Lanka: Government of Sri Lanka and United Nations.

Haigh, R., Amaratunga, D., Baldry, D., Pathirage, C. & Thurairajah, N. (2009). *ISLAND – Inspiring Sri Lankan Renewal and Development. RICS Research*. Salford, UK: University of Salford.

Halvorson, S. J. & Hamilton, J. P. (2010). In the Aftermath of the Qa'yamat: The Kashmir Earthquake Disaster in Northern Pakistan. *Disasters*, 34, 184–204.

James Lee Witt Associates (2005). *Building Back Better and Safer: Private Sector Summit on Post-tsunami Reconstruction*. Washington, DC: James Lee Witt Associates.

Johnson, C., Lizarralde, G. & Davidson, C. H. (2006). A Systems View of Temporary Housing Projects in Post-disaster Reconstruction. *Construction Management & Economics*, 24, 367–378.

Kennedy, J. (2009). Disaster Mitigation Lessons from "Build Back Better" Following the 26 December 2004 Tsunamis. In: Ashmore, J., Babister, E., Kelman, I. & Zarins, J. (eds) *Water and Urban Development Paradigms*. London: Taylor & Francis.

Kennedy, J., Ashmore, J., Babister, E. & Kelman, I. (2008). The Meaning of "Build Back Better": Evidence from Post-tsunami Aceh and Sri Lanka. *Journal of Contingencies & Crisis Management*, 16, 24–36.

Khasalamwa, S. (2009). Is "Build Back Better" a Response to Vulnerability? Analysis of the Post-tsunami Humanitarian Interventions in Sri Lanka. *Norwegian Journal of Geography*, 63, 73–88.

Lyons, M. (2009). Building Back Better: The Large-scale Impact of Small-scale Approaches to Reconstruction. *World Development*, 37, 385–398.

Mannakkara, S. & Wilkinson, S. (2012a). Build Back Better Principles for Economic Recovery: The Victorian Bushfires Case Study. *Journal of Business Continuity and Emergency Planning*, 6, 164–173.

Mannakkara, S. & Wilkinson, S. (2012b). Build Back Better Principles for Land-use Planning. *Urban Design and Planning*, 166, 288–295.

Meese III, E., Butler, S. M. & Holmes, K. R. (2005). *From Tragedy to Triumph: Principled Solutions for Rebuilding Lives and Communities. Heritage Special Report.* Washington, DC: Heritage Foundation.

Middleton, D. (2008). Habitability of Homes after a Disaster. *4th International i-REC Conference on Building Resilience: Achieving Effective Post-disaster Reconstruction.* Christchurch, New Zealand: International Council for Research and Innovation in Building and Construction.

Monday, J. L. (2002). Building Back Better: Creating a Sustainable Community after Disaster. *Natural Hazards Informer*, 3. Available: https://hazards.colorado.edu/archive/publications/informer/infrmr3/informer3b.htm [Accessed 9 August 2018].

Okuda, K., Ohashi, M. & Hori, M. (2011). On the Studies of the Disaster Recovery and the Business Continuity Planning for Private Sector Caused by Great East Japan Earthquake. In: Cruz-Cunha, M. M., Varajão, J., Powell, P. & Martinho, R. (eds) *ENTERprise Information Systems.* Berlin, Germany: Springer.

Olshansky, R. B. (2005). How Do Communities Recover from Disaster? A Review of Current Knowledge and an Agenda for Future Research. *46th Annual Conference of the Association of Collegiate Schools of Planning.* Kansas City, MO.

Ozcevik, O., Turk, S., Tas, E., Yaman, H. & Beygo, C. (2009). Flagship Regeneration Project as a Tool for Post-disaster Recovery Planning: The Zeytinburnu Case. *Disasters*, 33, 180–202.

Pathiraja, M. & Tombesi, P. (2009). Towards a More "Robust" Technology? Capacity Building in Post-tsunami Sri Lanka. *Disaster Prevention and Management*, 18, 55–65.

Red Cross (2010). *World Disasters Report 2010 – Focus on Urban Risk.* Geneva, Switzerland: International Federation of Red Cross and Red Crescent Societies.

Ruwanpura, K. N. (2009). Putting Houses in Place: Rebuilding Communities in Post-tsunami Sri Lanka. *Disasters*, 33, 436–456.

Tas, M. (2010). Study on Permanent Housing Production after 1999 Earthquake in Kocaeli (Turkey). *Disaster Prevention and Management*, 19, 6–19.

TEC (2007). *Consolidated Lessons and Recommendations from the TEC Synthesis Report and TEC Thematic Evaluations.* London: Tsunami Evaluation Coalition.

Twigg, J. (2007). *Characteristics of a Disaster-resilient Community – a Guidance Note.* London: DFID Disaster Risk Reduction Interagency Coordination Group.

UNDRO (1982). *Shelter after Disaster: Guidelines for Assistance.* New York: United Nations Disaster Relief Organisation.

UNEP (2008). "Disasters and Conflicts". *UN Environment.* Available: https://www.unenvironment.org/explore-topics/disasters-conflicts [Accessed 9 August 2018].

VBRRA (2010). *15 Month Report.* Melbourne, Australia: Victorian Bushfire Reconstruction and Recovery Authority.

VBRRA (2011). *24 Month Report.* Melbourne, Australia: Victorian Bushfire Reconstruction and Recovery Authority.

Winchester, P. (2000). Cyclone Mitigation, Resource Allocation and Post-disaster Reconstruction in South India: Lessons from Two Decades of Research. *Disasters*, 24, 18–37.

3 Structural Resilience

Introduction

A primary reason for extensive damage from natural disasters is the inadequate structural capacity of the built environment. The damage from the 2004 Indian Ocean Tsunami (DN and PA, 2008) and the 2009 Samoan Tsunami (Bird et al., 2011) was partly due to insufficient consideration of coastal risks in the design of structures. The destruction from the Kashmir Earthquake in Pakistan was again traced to non-earthquake-resistant building design and construction in earthquake-prone areas (Halvorson and Hamilton, 2010). Similar cases were apparent in the 2010 Haiti Earthquake and the 1995 Kobe Earthquake disasters (Ellsworth, 1995). Destruction in the built environment from disasters has been attributed to the lack of structural resilience, including incomplete and inaccurate assessments of hazards, incompatibilities between structural designs and hazard levels, lack of consideration of risks in town planning, and neglected building codes and regulations (Dias et al., 2006, Halvorson and Hamilton, 2010, Mora and Keipi, 2006). These factors result in a vulnerable built environment. As part of risk reduction, the need for housing and infrastructure to be built back better becomes essential to reduce future vulnerabilities and increase resilience. In addressing structural resilience as one of the key elements for creating a better post-event environment, governments can ensure communities return to more hazard-resistant infrastructure and safer buildings. In encouraging structural resilience, this chapter explores the adoption of building codes and regulation and the impact of cost and time-related factors on rebuilding better. In addition, the quality versus speed dilemma in the context of structural resilience and BBB is explained. BBB Indicators for improving structural resilience, which serve as practical recommendations for implementation are presented, along with case study examples demonstrating Building Back Better through improved structural resilience.

Structural damage and the need for Building Back Better

Clinton (2006) said that a key test of a successful recovery effort is whether it leaves survivors less vulnerable to natural hazards. Build Back Better advocates that reconstruction provides an opportunity to incorporate risk reduction measures while damaged structures are rebuilt to eliminate pre-existing vulnerabilities and increase resilience to future hazards (Palliyaguru and Amaratunga, 2008, Mercer, 2010). The importance of reviewing and changing building designs and codes to improve the structural integrity of buildings and infrastructure following a disaster is widely understood (Meigh, 2009, Halvorson and Hamilton, 2010, McCurry, 2011), but is less frequently attained successfully in practice due to a range of common issues. Poor regulatory powers and the lack of strict enforcement can lead to building code changes being disregarded, resulting in sub-standard structures in the rebuild (DN and PA, 2008, Asian Development Bank et al., 2005). When the Indian Ocean Tsunami struck, enforcement of building codes was mainly restricted to urban and suburban areas in Sri Lanka. The rural and coastal areas were the main victims of the disaster, where the lack of strict structural standards resulted in magnified damage (Palliyaguru et al., 2010, Pathiraja and Tombesi, 2009). Extra costs incurred by adopting new technologies and materials to improve structural resilience also discourage compliance with new building codes (Batteate, 2006, Kijewski-Correa and Taflanidis, 2012).

The findings of Egbelakin et al. (2011) indicate that confusion created by contradictory information, and the perception of building owners about the risks of another disaster event in the near future contributed towards the scepticism about building strengthening to reduce future risks. As soon as a disaster strikes, it is common to see recovery efforts addressing only the just-experienced hazard, which can exacerbate vulnerabilities to other hazards (Kennedy, 2009).

Time pressures in the recovery process with expectations for fast results also largely contribute to hasty design and construction in the absence of well-thought out building codes and hazard assessments (Kennedy et al., 2008). The last factor that contributes to poor structural integrity is the quality of workmanship during construction. It is common practice for non-governmental organisations, imported tradespeople and home-owners to assist with reconstruction projects, leading to inconsistent quality in the rebuild (Khasalamwa, 2009, Boano, 2009, Pathiraja and Tombesi, 2009).

The experiences of post disaster reconstruction efforts worldwide have provided lessons which can be adopted when implementing structural changes to Build Back Better. Building codes and regulations have been a topic of interest post disaster, particularly as the codes and regulations relate to Building Back Better. For instance, advocates of hazard-based building regulations believe there is a need to use multi-hazard assessments in areas chosen for redevelopment and reconstruction (Batteate, 2006, Haigh et al., 2009, FEMA, 2000, UNISDR, 2005). In addition, researchers have claimed that a strong legal framework assists the adoption of better building codes and regulations and ensures that structural changes improve the built environment for the affected community (Iglesias et al., 2009, Mora and Keipi, 2006, Clinton, 2006). Ways of reducing risk can be promoted through community programmes, such that the community and stakeholders are educated about adopting community-inclusive risk reduction and better building practices using changed regulations (Ikeda et al., 2007, Reddy, 2000).

Structural code changes post disaster are common. However, there are cost and time implications in improving structural resilience post disaster. Funding needs to be made available to cover extra costs for structural improvements (DN and PA, 2008), and the adoption of new building regulations needs to be promoted alongside appropriate incentives to improve, such as tax or insurance reductions (Bakir, 2004, Edwards, 2010). As recovery takes time, pre-event planning creates an opportunity to embed structural changes before hazard events and to reduce the need to spend time in making structural changes post disaster (Olshansky, 2005).

One of the main issues with speed of recovery is the impact on quality. Grewal believes that quality should not be compromised for speed (Grewal, 2006). Recovery done quickly often produces a built environment that requires ongoing replacement, repair and maintenance. A focus on quality products requires inspections during construction, and qualified authorities should make these inspections with a specific focus on quality (Lewis, 2003). If owner-built reconstruction is used during recovery, houses will require professional supervision (Pathiraja and Tombesi, 2009). In order to focus on quality, training of stakeholders, including the affected community and others involved in the design and construction, using new design and construction information required for the rebuild is needed (Lloyd-Jones, 2006).

Illustrations of Building Back Better: successes and opportunities

Case study: improving structural resilience in Fijian schools after 2016 Tropical Cyclone Winston

Tropical Cyclone Winston (TC Winston) struck Fiji in February 2016, impacting 62% of Fiji's total population. TC Winston was one of the strongest recorded tropical storms in the southern hemisphere, with wind gusts reaching 330 kilometres per hour, and it was classed as a Category 5 storm. The cyclone tracked along Fiji, impacting numerous islands, with Koro Island being one of the worst-hit. Entire communities were destroyed, along with 88 health clinics, 30,369 houses and 495 schools damaged. The estimated cost of destruction was 731 Fijian dollars. The large number of schools damaged led to the rebuilding of schools being a major focus of the reconstruction effort.

The Fijian Government developed the Disaster Recovery Framework (DRF) to guide the recovery effort with a vision for a "Stronger and More Resilient Fiji". The guiding principles in the DRF included:

- Building Back Better – to build to a higher construction standard to reduce vulnerability and improve living conditions;
- Inclusive – to include all sectors of society and work with public, private and non-profit organisations;
- Pro-poor – to address the most vulnerable individuals and improve their access to financial services;
- Building resilient communities and institutions – to include disaster risk management and climate change adaptations at the core of government documents to achieve risk-improved development and review disaster risk management policies.

Vunikavikaloa Arya Primary School reconstruction

The intent of this project was to provide a platform for the Fijian Red Cross Society (FRCS) to access Fijian-Indian communities that it may not have been able to reach. Thus, the selection and adoption of Vunikavikaloa Arya School in Nawala that was significantly damaged by Tropical Cyclone Winston on 20 February 2016 was both thoughtful and considered. However, there were many constraints on this project from the various stakeholders.

The New Zealand Red Cross (NZRC) had issues regarding the extent of damage, community engagement and the design of the new school. The NZRC had questions about the timeline and costs for the work and how it would be implemented, and about the risks and assumptions involved in such a project.

The FRCS wanted to get access to this community by building the school, and so viewed the building as the start rather than the conclusion of a relationship with the wider community of approximately five villages that the school served. It had a desire to adopt a participatory approach, including capacity-building, to rebuild the school while also up-skilling the community to rebuild their own houses. Volunteers from the local community, and potentially from the nearby Ra High School's trade training course, would be the core construction team in this case.

The Ministry of Education (MoE) had standards and codes that needed to be adhered to, and it was not completely clear what they might be with this project, but certainly it was anticipated to include details such as inside finishes. There was also the question of whether the school would or should be an evacuation centre, which would require a higher level of wind loading design.

The Construction Implementation Unit (CIU) which was overseeing the construction of all schools on behalf of the government required the Vunikavikaloa Arya School rebuild to be completed on time and to "accepted" standards. Consequently, it expected a comparable level of quality of finished construction as would be expected from other contractors. It was looking to have all 220 schools it was working on completed by end of January 2017, in time for the start of school.

The Fiji Institute of Engineers (FIE) was involved in ensuring that the school design was signed off by a Fijian Registered Engineer. There was discussion among the engineers about the wind load design levels. Current Gazetted Fijian Codes from 1990 are based on outdated New Zealand and Australian codes (for example, NZS4203), and there are various versions of recent codes in circulation based on the ANZS 1170 series. However, the FIE is promoting design to cope with Category 5 cyclones (taken as 66 metres per second wind design speed), with a design wind speed of 75–77 metres per second for evacuation centres, which results in a difference in load of 36% between the two designs.

The teaching staff at the school were desperately looking for their classrooms to be back and operational. Tents were built as temporary classrooms, which worked well, but were difficult to teach in due to practical issues as well as the heat. They needed to be shifted regularly in the rainy season due to the muddy ground conditions. The teachers were concerned that the school roll had dropped from 250 to 230 students, and wanted the school to be restored as soon as possible.

The building context in Fiji was characterised by a shortage of materials and skilled tradespeople, along with high material costs. The shortage of materials was being addressed by importing materials, especially timber, but there was still a shortage of building strapping material.

Rebuilding Vunikavikaloa Arya Primary School was a challenging task, especially in attempting to take a participatory approach using local builders and community members while ensuring compliance with design levels above the usual construction practices and building codes as required by the MoE and CIU.

The Fijian Institute of Engineers set up an "Adopt-a-School" programme encouraging foreign governments, donor agencies, companies, community groups, sporting bodies and individuals to adopt schools and lead the rebuilding process. Chand Engineering Consultants adopted Vunikavikaloa School and prepared a structural report on repairs and costing for the school, while the FRCS adopted the school to complete the repair and rebuilding. The FRCS chose to use FRAMECAD technology developed in New Zealand for the school rebuild, which is a rapid construction method using light-gauge steel framing that can be easily screwed together to form trusses. The technology allows high-quality, fast construction with simple construction methods and training, which allowed local people to become involved in the rebuild. The FRCS adopting FRAMECAD technology to undertake the school rebuild was a "Build Back Better" solution which addressed all the issues associated with this project. The design met Category 5 loading standards, the construction had high quality assurance, and the materials and skills shortages were overcome, along with cost and time savings. The school was completed on time, under budget and using local people, which has been a source of pride for the school and community.

Case study: improving structural resilience in Sri Lanka after the 2004 Indian Ocean Tsunami

The tsunami waves resulting from the 9.0 magnitude earthquake which occurred off the coast of Sumatra, Indonesia on 26 December 2004 impacted 14 countries, including Sri Lanka, which suffered substantial damage. Thirteen out of 25 coastal districts in the east and south of Sri Lanka were affected, with 35,322 lives lost, 516,150 people displaced and approximate direct losses of US$1 billion. At the time of the tsunami, Sri Lanka did not have strictly enforced building codes for residential developments, and had complicated and time-consuming permit procedures which were largely ignored, leading to vulnerable settlements in high-risk areas such as the coastal belt. Following the tsunami, all the rebuilding work and livelihood support were handed over to the NGO sector, and the Task Force to Rebuild the Nation (TAFREN) and the subsequent body, the Reconstruction and Development Agency (RADA), were established to coordinate and assist stakeholders in the rebuilding phase.

It is usual practice to see building codes and regulations revised following a disaster, but reconstruction in Sri Lanka was primarily based on relocating communities away from high-tsunami risk areas, with less focus on improving structural designs. This led to the construction of sub-standard vulnerable structures after the tsunami. Guidelines for building in areas at risk from natural disasters was published in October 2005 to aid the Sri Lankan tsunami rebuild, containing general non-specific principles for single-storey construction to resist earthquake, cyclone and flood impacts, but the guidelines were not followed because of the rush to rebuild as fast as possible. In terms of structural resilience, the tsunami rebuild in Sri Lanka was not successful, and highlighted the need for regulatory authorities to assert control over the structural integrity of buildings through the use of building codes and regulations. As a result of the tsunami experience, building guidelines and codes have been developed to address different hazards.

A common problem that resulted from the reactive relocation strategy was that people were being relocated to areas prone to other hazards, such as flooding, without adequate counter-measures in building designs. This increased their vulnerability, contrary to what was intended.

The reconstruction strategy adopted allowed housing reconstruction to be either donor-driven or owner-driven. Donor-driven construction involved houses being built and supplied to tsunami victims by donors (primarily NGOs), while owner-driven construction involved cash being supplied to home-owners to carry out the rebuild themselves. Although some NGOs maintained high quality standards, the lack of awareness of some international NGOs about local conditions contributed to the construction of inappropriately built structures.

36 *Structural Resilience*

Owner-driven housing programmes were favoured by the community, as the community felt good about being involved directly. However, home-owners often took the opportunity to modify their houses, for instance adding extra storeys on houses designed for one-storey loads, which compromised structural integrity. In order to implement participatory rebuilding successfully, it is important for construction to be supervised by trained personnel and inspected by regulative authorities to assure compliance.

Inadequate funding was also an issue which impacted the rebuild in Sri Lanka. The main source of funding was from donors, which were plentiful, but the use and distribution of donations was impeded by donor agendas, political influence and corruption. Donor funding was also short-term, and therefore insufficient to sustain construction projects over time. Long-term funding was needed.

Lengthy and time-consuming permit procedures discouraged the adoption of proper structural improvements measures in the rebuild in Sri Lanka. NGOs were working under pressure to meet deadlines to satisfy beneficiary expectations, which drove them to focus on speed and thus overlooked local permit protocols, which resulted in poor-quality construction in some instances. Simplification and clarification of procedures and institutional arrangements would help speed up and maintain the quality of reconstruction efforts.

Case study: improving structural resilience in Australian towns after the 2009 Victorian Bushfires

The Victorian Bushfires took place on the 7 of February 2009, when fires swept through 78 communities in the state of Victoria; 173 lives were lost, and more than 430,000 hectares of land, 2,000 properties, 55 businesses, 3,550 agricultural facilities, 70 national parks, 950 local parks and 467 cultural sites were destroyed. Some of the areas affected by the Victorian bushfires had not been declared as bushfire-prone before the event. The Australian building code for bushfire-prone areas was still in the process of being updated with the findings from the 2003 Canberra Bushfires when the 2009 fires occurred. The absence of accurate up-to-date mapping, land-use planning and construction regulations contributed towards worsening the impact of the fires.

Shortly after the bushfires, a revised edition of the Australian Standard for construction of homes in bushfire-prone areas (AS:3959) was released. The revised code introduced more refined bushfire risk levels and contained more stringent design and construction specifications for better protection. Compliance was enforced through permit procedures.

The changes to the code were not easy to implement at first. Some products required by the revised code for higher-risk areas were not available on the market yet. The code was going through changes and updates during the rebuild which builders found difficult to follow. The cost increases that resulted from the code changes were also problematic for home-owners. For example, a house in the highest-risk zone (classified as BAL FZ) cost A$70,000–100,000 extra to rebuild to standard. The housing rebuild was driven by the availability of insurance. Houses that now had higher risk levels as a result of the code changes were unable to afford the upgrades as insurance was available only to rebuild to the pre-existing standard. The Victorian Bushfire Appeal Fund provided grants for construction, but the lack of regulation concerning how the money was to be utilised resulted in people spending the money without saving enough for construction.

The success of implementing improved structural regulations was also hampered by the extra time required to make necessary improvements, which was viewed as inconvenient in an environment where speed was crucial. In instances where special materials specified in the new regulations were not available, people started building without the proper materials. The ongoing code changes also caused delays. However, the provision of comfortable temporary accommodation allowed most affected people to remain patient and move into well-designed, safe homes. It is necessary to be aware of over-reliance on temporary accommodation, as some people began to remain in temporary accommodation over long periods of time without focusing on rebuilding.

Another issue which affected the rebuild was the difficulty in getting builders and tradespeople for the rebuild due to the remoteness of the bushfire-impacted areas. This led to many home-owners opting for owner-building and self-managing the rebuilding of their homes. The Rebuilding Advisory Service set up by the Victorian Bushfire Reconstruction and Recovery Authority was valuable for these home-owners to get advice and ensure that the rebuilding work was being done correctly. Since permit procedures were relaxed to speed up the rebuilding process, and quality assurance of buildings was the responsibility of the builder.

The Victorian Bushfires rebuild effort provided a good example of Building Back Better, where the importance of improving structural capacity for improved resilience was understood and displayed through revising and enforcing appropriate building codes. The rebuild effort also displayed some critical issues which impacted successful implementation of this BBB Principle of Structural Resilience which serve as lessons. It is necessary to consider the availability of resources required for any building code changes in the post disaster period. Building code revisions need to be practical and easy to implement. It is important to educate stakeholders about building code updates to facilitate adoption. The cost of structural improvements needs to be considered, and implementation needs to be assisted through long-term funding arrangements such as grants or low-interest loans, as well as the provision of monetary incentives to promote the adoption of structural changes, or restricting costly construction on high-risk lands. Proposed improvements for structural designs need to be within manageable and realistic cost and time limitations to ensure compliance. The bushfires experience showed that providing comfortable transitional homes which can house affected people for a considerable period of time can reduce time pressures, allowing structural changes to be made properly. The focus should, however, remain on permanent construction, and long-term reliance on temporary accommodation must be avoided. The use of skilled builders can also assist with speed and quality of reconstruction. It is therefore necessary to have a strategy to attract skilled builders for reconstruction. Providing support through a rebuilding advisory service is also beneficial to assist home-owners, ensure quality and avoid complications.

Case study: improving structural resilience of agricultural businesses in Gaza

Gaza, Palestine was subject to ongoing military assault for seven weeks in July 2014 by land, sea and air. At least 2,145 people were killed and over 60,000 homes were damaged or destroyed. The conflict created a scarcity of water, energy, food and shelter, while the agriculture industry in particular suffered heavily. Rapid damage and loss assessments conducted in 29 locations showed extensive damage to crop production, poultry farmers, livestock farms and fisheries amounting to nearly US$23 million in damage and losses.

With the agricultural business sector suffering considerable damage, one of the focuses of the post disaster reconstruction and recovery effort was to rehabilitate affected agribusinesses. Improving structural resilience was the most practical response for building resilience, as relocation to safer areas was not deemed a viable option. Improving structural resilience for the businesses meant improving the resilience of their physical assets through repairing damage incurred along with introducing technological improvements to the businesses. Greenhouse farmers took the opportunity to strengthen drainage systems inside the greenhouses, install on-site water storage tanks and adopt the use of rainwater harvesting. Poultry farmers installed more efficient heating systems, better humidity control systems and desalinisation systems for water. Livestock and dairy farmers opted for more secure and cost-effective barns, including modern technology such as steel feeders, mechanical drinking and isolation units for lambs, as well as better draining and ventilation for barns and water harvesting systems.

The Gaza case shows that improving the resilience of physical assets can assist business recovery as well by reducing risks posed to businesses through natural and man-made hazards.

Case study: improving structural resilience of tourism businesses in the Cook Islands

The Cook Islands is comprised of 15 small islands in the Southwest Pacific Ocean. Its capital city and largest island is Rarotonga. Its climate is dominated by the extensive and persistent South Pacific Convergence Zone (SPCZ), which has a major influence on the weather. The intensity and position of the SPCZ coupled with the natural geography and landscape of the islands have contributed to regular cyclones striking and causing devastation to the small nation, including deadly calamities such as Cyclone Martin in October–November 1997 and five destructive cyclones in February–March 2005. The cyclones have caused over NZ$20 million worth of damages to the country, the majority of which was evident in Rarotonga.

Tourism is the most dominant economic sector in the Cook Islands, providing over 60% of the national GDP. The main attraction in the Cook Islands is its natural assets such as beaches and natural scenery. Tourism infrastructure such as hotels, resorts, restaurants and other related businesses have been established in Rarotonga to support industry.

Overall, tourism providers had sturdy, resilient structures that were able to withstand the cyclones encountered. Businesses adopted traditional preparatory practices such as placing shutters on windows and doors, tying roofs down where applicable, and knowing where evacuation routes and sites were. Accommodation businesses had structures constructed primarily of concrete and steel. The popular small to medium-sized tourist businesses had portable designs which allowed easy extraction of assets for evacuation during cyclones. The resilience of small businesses depended on preserving the main assets needed to operate the business, with less focus on building damage. Businesses took the initiative to modify their landscapes and replace roofs with lighter alternatives. For example, some businesses changed their steel roofs to aluminium roofs and all screws to stainless steel, to counteract rusting. Other businesses installed seawalls and groynes to withstand forces imposed by the sea. Businesses that had sustained serious damage from previous cyclones had redesigned their buildings to include concrete floors and quick-release measures to allow certain components and goods to be moved in the event of a cyclone.

Businesses in Rarotonga did not follow specific building codes or regulations. Larger businesses tried to comply with New Zealand standards, although they were not very applicable to the Cook Islands. Other businesses followed a combination of New Zealand standards and local best practices. Tourism businesses suffered from the heavy cost of insurance due to businesses being located by the sea, and therefore considered as high-risk for tourism purposes. Large businesses were insured, while smaller non-accommodation tourism businesses remained largely uninsured.

Tourism businesses in Rarotonga were not interested in relocating away from the prime beach-side locations, and therefore have opted to minimise risks through improved structural resilience and other risk reduction measures such as the installation of seawalls and groynes, and comprehensive safety and evacuation plans and procedures.

Improving structural resilience for Building Back Better

The implementation of structural changes during post disaster reconstruction to improve structural resilience can be achieved more successfully by adopting BBB Principles related to structural improvements. The BBB Principles for structural changes were categorised under Building Codes and Regulations, Cost and Time and Quality based on analysis of existing literature.

BBB practices for improving structural resilience in order to Build Back Better are universally applicable despite legal, political, social, cultural and administrative differences. Rebuilding has to be led by accurate, up-to-date hazard-based building regulations using multi-hazard assessments which take disaster risk reduction and climate change into account. The adoption of building codes and guidelines should be supported by a strong legal framework in place. The community and building practitioners need to be educated about the importance of following building regulations. In the event of structural improvements executed to improve structural resilience, long-term funding mechanisms such as grants, low-interest loans and insurance have to be arranged to cover extra costs resulting from the improvements to facilitate adoption. Providing incentives to promote adoption of building codes is also a strategy that can be considered for BBB. Quality assurance is a key factor in ensuring structural resilience during the rebuild, and this needs to be supported though methods such as regular inspections if practical, and the use of trained skilled builders. Ensuring structural resilience in the rebuild also requires training to be provided to building practitioners and stakeholders on modified structural requirements and new regulations.

BBB includes considering alternatives when improving structural resilience becomes impractical. For example, costly and timely construction on high-risk lands can be restricted by coordinating with planners. It is necessary to consider options such as providing comfortable short-term transitional accommodation to relieve time pressures on reconstruction. This can allow extra time to perform thorough hazard assessments, conduct stakeholder training programmes, revise building codes, consult the community and complete necessary activities to provide the best possible solutions for making communities safer. The quality of the rebuild is enhanced by using skilled builders. Incentives such as accommodation options and subsidised resources can be provided to attract skilled builders to take part in the rebuild. Establishing a database of potential builders in the pre-disaster period who would be interested in taking part in reconstruction activities would also make the process more efficient. Having a pre-planned strategy is key to a successful reconstruction process.

This chapter demonstrates the need to consider wider implications of making communities safer through improving structural resilience in the built environment, and encourages recovery and reconstruction planners and implementers to consider and address a wide range of issues before commencing the rebuild in order to build back better.

BBB Indicators for Structural Resilience

Indicators or best practices for Building Back Better for improving Structural Resilience were developed based on case study research findings and international examples. These Indicators are listed in Table 3.1, grouped under Building Codes, Cost and Time, and Quality. The Indicators serve as a practical guide to direct stakeholders involved in post disaster activities to understand the elements that need to be considered when planning and implementing reconstruction and recovery programmes in order to Build Back Better.

Table 3.1 Build Back Better Indicators for improving Structural Resilience

Building Codes	Accurate, up-to-date understanding of all hazards affecting physical assets and resulting disaster risks and vulnerabilities
	Legislated structural codes based on up-to-date multi-hazard assessments with design and construction specifications aligned with local resource availability and affordability
	Building codes and regulations that incorporate traditional technologies and are aligned with local knowledge and skills
	Education provided to stakeholders on post disaster building regulation changes prior to commencing rebuilding work
Cost and Time	Long-term funding mechanisms in place to fund extra costs incurred for structural improvements
	Incentives planned and in place to promote adoption of structural improvements to homes, buildings and infrastructure
	Unaffordable construction on high-risk lands restricted
Quality	Redundancies and transitional arrangements in place to relieve pressures on fast and reactive rebuilding (transitional accommodation, alternative travel routes etc.) and implement well-planned rebuild projects
	Practical, uncomplicated and efficient quality assurance measures in place appropriate for post disaster environments
	Incentives or special arrangements (e.g. alliance or public-private partnerships) planned and in place to attract skilled certified builders to meet reconstruction demands
	Provide professional supervision for owner-building of homes for quality assurance
	Rebuilding advisory service with rebuilding advisors in place to support, educate and assist home-owners needing to rebuild

References

Asian Development Bank, Japan Bank for International Cooperation & World Bank (2005). Preliminary Damage and Needs Assessment. In: *Sri Lanka 2005 Post-Tsunami Recovery Programme*. Colombo, Sri Lanka: Asian Development Bank, Japan Bank for International Cooperation and World Bank.

Bakir, P. G. (2004). Proposal of a National Mitigation Strategy against Earthquakes in Turkey. *Natural Hazards*, 33, 405–425.

Batteate, C. (2006). Urban Disaster Risk Reduction and Regeneration Planning: An Overview. *Focus: Journal of the City and Regional Planning Department*, 3, 11–17.

Bird, D. K., Chague-Goff, C. & Gero, A. (2011). Human Response to Extreme Events: A Review of Three Post-tsunami Disaster Case Studies. *Australian Geographer*, 42, 225–239.

Boano, C. (2009). Housing Anxiety and Multiple Geographies in Post-tsunami Sri Lanka. *Disasters*, 33, 762–785.

Clinton, W. J. (2006). *Lessons Learned from Tsunami Recovery: Key Propositions for Building Back Better*. New York: Office of the UN Secretary-General's Special Envoy for Tsunami Recovery.

Dias, P., Dissanayake, R. & Chandratilake, R. (2006). Lessons Learned from Tsunami Damage in Sri Lanka. *Proceedings of ICE Civil Engineering*, 159, 74–81.

DN & PA (2008). *Disaster and Development in South Asia: Connects and Disconnects. South Asia Disaster Report.* Colombo, Sri Lanka: Duryog Nivaran and Practical Action.

Edwards, W. (2010). State Building Codes Remain Inadequate Five Years after Hurricane Katrina Hit. *National Underwriter/P&C*, 114, 28.

Egbelakin, T. K., Wilkinson, S., Potangaroa, R. & Ingham, J. (2011). Challenges to Successful Seismic Retrofit Implementation: A Socio-behavioural Perspective. *Building Research & Information*, 39, 286–300.

Ellsworth, W. L. (1995). From California to Kobe. *Nature*, 373, 388.

FEMA (2000). *Rebuilding for a More Sustainable Future: An Operational Framework.* Washington, DC: Federal Emergency Management Agency.

Grewal, M. K. (2006). Sri Lanka – a Case Study. In: *Approaches to Equity in Post-tsunami Assistance.* Colombo, Sri Lanka: Office of the UN Secretary General's Special Envoy for Tsunami Recovery.

Haigh, R., Amaratunga, D., Baldry, D., Pathirage, C. & Thurairajah, N. (2009). *ISLAND – Inspiring Sri Lankan Renewal and Development. RICS RESEARCH.* Salford, UK: University of Salford.

Halvorson, S. J. & Hamilton, J. P. (2010). In the Aftermath of the Qa'yamat: The Kashmir Earthquake Disaster in Northern Pakistan. *Disasters*, 34, 184–204.

Iglesias, G., Arambepola, N. M. S. I. & Rattakul, B. (2009). Mainstreaming Disaster Risk Reduction into Local Governance. In: *National Symposium on Creating Disaster Free Safer Environment.* Colombo, Sri Lanka: National Building Research Organisation and Ministry of Disaster Management and Human Rights.

Ikeda, S., Sato, T. & Fukuzono, T. (2007). Towards an Integrated Management Framework for Emerging Disaster Risks in Japan. *Natural Hazards*, 44, 267–280.

Kennedy, J. (2009). Disaster Mitigation Lessons from "Build Back Better" Following the 26 December 2004 Tsunamis. In: Ashmore, J., Babister, E., Kelman, I. & Zarins, J. (eds) *Water and Urban Development Paradigms.* London: Taylor & Francis.

Kennedy, J., Ashmore, J., Babister, E. & Kelman, I. (2008). The Meaning of "Build Back Better": Evidence from Post-tsunami Aceh and Sri Lanka. *Journal of Contingencies & Crisis Management*, 16, 24–36.

Khasalamwa, S. (2009). Is "Build Back Better" a Response to Vulnerability? Analysis of the Post-tsunami Humanitarian Interventions in Sri Lanka. *Norwegian Journal of Geography*, 63, 73–88.

Kijewski-Correa, T. & Taflanidis, A. (2012). The Haitian Housing Dilemma: Can Sustainability and Hazard-resilience Be Achieved? *Bulletin of Earthquake Engineering*, 10, 765–771.

Lewis, J. (2003). Housing Construction in Earthquake-prone Places: Perspectives, Priorities and Projections for Development. *Australian Journal of Emergency Management*, 18, 35–44.

Lloyd-Jones, T. (2006). *Mind the Gap! Post-disaster Reconstruction and the Transition from Humanitarian Relief.* London: Royal Institution of Chartered Surveyors.

McCurry, J. (2011). How Past Japan Earthquakes Prepared Nation for Today's Historic Temblor. *Christian Science Monitor*, 11 March.

Meigh, D. (2009). Aceh Emergency Support for Irrigation – Building Back Better. *Proceedings of ICE Civil Engineering*, 162, 171–179.

Mercer, J. (2010). Disaster Risk Reduction or Climate Change Adaptation: Are We Reinventing the Wheel? *Journal of International Development*, 22, 247–264.

Mora, S. & Keipi, K. (2006). Disaster Risk Management in Development Projects: Models and Checklists. *Bulletin of Engineering Geology and the Environment*, 65, 155–165.

Olshansky, R. B. (2005). How Do Communities Recover from Disaster? A Review of Current Knowledge and an Agenda for Future Research. *46th Annual Conference of the Association of Collegiate Schools of Planning*. Kansas City, MO.

Palliyaguru, R. & Amaratunga, D. (2008). Managing Disaster Risks through Quality Infrastructure and Vice Versa: Post-disaster Infrastructure Reconstruction Practices. *Structural Survey*, 26, 426–434.

Palliyaguru, R., Amaratunga, D. & Haigh, R. (2010). Integration of "Disaster Risk Reduction" into Infrastructure Reconstruction Sector: Policy vs Practise Gaps. *International Journal of Disaster Resilience in the Built Environment*, 1, 277–296.

Pathiraja, M. & Tombesi, P. (2009). Towards a More "Robust" Technology? Capacity Building in Post-tsunami Sri Lanka. *Disaster Prevention and Management*, 18, 55–65.

Reddy, S. D. (2000). Factors Influencing the Incorporation of Hazard Mitigation During Recovery from Disaster. *Natural Hazards*, 22, 185–201.

UNISDR (2005). *Hyogo Framework for Action 2005–2015: Building the Resilience of Nations and Communities to Disasters*. Kobe, Japan: United Nations Office for Disaster Risk Reduction.

4 Multi-hazard-based Land-use Planning

Introduction

Disasters are a result of hazards and vulnerability (Guha-Sapir et al., 2004). Exposure to hazards is a significant cause for increased vulnerability (Quarantelli, 1987, Morrow, 1999). Major disasters such as Hurricane Katrina, the Kashmir Earthquake, the Indian Ocean Tsunami, the Samoan Tsunami as well as frequent small-scale disasters such as local flood events have shown that location plays a significant role in the level of destruction caused (Kates et al., 2006, GoSL, 2005, Halvorson and Hamilton, 2010, De, 2011). The recovery strategies and lessons learnt from these disasters have all pointed towards the importance of land-use planning strategies to minimise and mitigate disaster risks.

This chapter illustrates how multi-hazard-based land-use planning is a central part of reducing risk and Building Back Better. Resettlement is often used as a means of creating communities away from hazards, but this creates other problems for the communities, such as less access to previous livelihoods. This chapter examines how land-use planning and resettlement can be better aligned with BBB in the best interests of the affected communities.

Post disaster land-use planning

The occurrence of a major disaster event questions the safety of a community's location and whether the community should be relocated to a lower-risk area. The damage from the 2004 Indian Ocean Tsunami (DN and PA, 2008) and the 2009 Samoan Tsunami (Bird et al., 2011) was partly due to insufficient consideration of coastal risks in land-use planning. BBB advocates that during reconstruction, hazard-resistant structures are to be built with better consideration of land-use planning to minimise damage from future natural hazards (Kennedy et al., 2008, Palliyaguru and Amaratunga, 2008). Post disaster recovery after the Indian Ocean Tsunami and the Samoan Tsunami both resulted in the relocation of coastal communities further inland to prevent future impacts of coastal hazards (Kennedy et al., 2008, Potangaroa, 2009). The mandatory resettlement operations in Sri Lanka and Samoa were problematic due to the lack of consideration given to the lifestyles of the local people, which led to the loss of their sea-dependent livelihoods, dissatisfaction with the new settlements and illegal return of people to their original coastal lands (Kennedy et al., 2008, Frerks and Klem, 2005, Birkmann and Fernando, 2008). A recurring issue with relocation is the focus on moving communities away from a certain hazard, resulting in exposing communities to new unanticipated hazards (Mora and Keipi, 2006, Red Cross, 2010). Oliver-Smith (1991) recommends attempting to rehabilitate original sites first, with resettlement considered as a last resort.

Baradan (2006) and Palliyaguru et al. (2010) have stressed the importance of developing risk-based land-use plans. Effective land-use planning avoids the need to adopt costly structural design solutions to manage risks (Iglesias et al., 2009, James Lee Witt Associates, 2005). For example, land-use planning was a key risk reduction measure adopted in the 2009 Victorian Bushfires rebuild. The improved risk-based land-use plans were useful in determining appropriate planning and building requirements (DPCD [Department of Planning and Community Development], 2013, 2009 Victorian Bushfires Royal Commission, 2010). A resettlement strategy called the "bushfire buy-back scheme", where the Victorian government bought back high-risk lands, was also introduced to prevent future residential building on lands classified as having very high bushfire risks and to encourage existing home owners to voluntarily relocate to lower-risk lands.

Carrying out multi-hazard assessments of current land sites and possible new land sites and creating risk zone maps which are used to determine appropriate land-uses and new planning and building regulations is recommended practice (Haigh et al., 2009, Baradan, 2006, Iglesias et al., 2009). The risk zone maps should be legislated and included in council development plans and approval permit procedures to ensure compliance (Iglesias et al., 2009, Glavovic, 2010, DMC [Disaster Management Centre] et al., 2011). Examples such as Taiwan's Mitigation Plans, the Philippines Municipal Maps and the Christchurch City

Plan in New Zealand following the Canterbury Earthquakes display successful application of BBB measures to create safer developments (Iglesias et al., 2009, Batteate, 2005, CERA, 2011).

Well-intended land-use planning measures can fail due to the lack of knowledge and awareness of local people who do not conform to new regulations, and the lack of experience and knowledge of local governing authorities who do not enforce new regulations (Kennedy et al., 2008, DFID, 2004). Olsen et al. (2005), Glavovic (2010) and DN and PA (2008) encourage educating communities about risks and the importance of risk reduction measures and engaging them in collective risk reduction efforts as part of BBB. The Participatory Flood Risk Communication Support System (Pafrics) developed in Japan to educate locals and other stakeholders, including NGOs and local governments, on flood risks and risk management strategies is an example of a participatory tool (Ikeda et al., 2007).

Resettlement has often created many problems for communities. Regardless of the advantages of improved safety, many communities do not wish to move from their original locations due to generational and cultural attachments with the land and their community, and the impact of relocation on their livelihoods and lifestyles. Therefore, relocation should be a last resort option (Oliver-Smith, 1991), and must only be opted for if rebuilding in the original location is not feasible and resettling can provide improved safety, along with adequate infrastructure, business and livelihood opportunities, and educational, health and recreational facilities.

Illustrations of Building Back Better successes and opportunities

Case study: 2009 Samoa Earthquake and Tsunami: Shall I stay or shall I go?

The 2009 Samoa Earthquake and Tsunami took place on 29 September 2009 as a result of a submarine earthquake with a magnitude of 8.1. Tsunami waves generated from the earthquake reached Samoa, American Samoa and Tonga, resulting in widespread destruction and loss of life. In Samoa, approximately 3,000 people were displaced and the cost of damages was estimated at US$147 million.

The Samoan Government looked at three options following the tsunami rebuild:

- Option 1 – Encourage relocation outside the tsunami inundation zone as a "Build Back Better" strategy. This would ensure safer communities as a result of being relocated to lower-risk areas.
- Option 2 – Allow families and affected households to make their own decisions between relocation or rebuilding. This option would require infrastructure at both locations, such as water supply, electrical supply and roading, in addition to the location of educational buildings, which could be difficult practically. This would also mean that those opting to remain along the coast would still be at risk.

- Option 3 – Allow rebuilding, but without any provision of services. This was the most economical option for the government, but provided the least protection for affected families.

The Samoan recovery effort considered how improvements could be made to the communities during the rebuild amid practical difficulties. The safest option would have been relocating the communities. However, the reality was that many households had strong long-term attachments to their lands, such as having relatives buried in those locations, which made relocation a complicated decision.

The Samoan experience showed that multi-hazard-based land-use planning needs to consider the tangible as well as the intangible aspects of the situation involved. These issues cannot and should not be ignored, but instead must be folded into BBB analyses if realistic successful long-term solutions are to be realised.

Case study: land-use planning in Sri Lanka after the 2004 Indian Ocean Tsunami

The primary risk reduction strategy implemented in Sri Lanka following the 2004 Indian Ocean Tsunami was the application of a "coastal buffer zone" banning construction along the coastal strip, which was considered a high-tsunami risk zone. People who previously lived within the buffer zone areas were relocated. The buffer zone resulted in two reconstruction policies: (1) buildings which were previously within the buffer zone were to relocate outside the buffer zone, and (2) buildings which were outside the buffer zone were to be rebuilt in situ.

The mandatory decision to relocate caused problems for recovery in Sri Lanka, and Boano (2009) described the buffer zone as "the single greatest barrier to progress" in post-tsunami reconstruction. One of the issues with relocation was the scarcity of suitable available lands for relocation. There was not enough available state land, and lands that were available had problems with water availability, infrastructure and illegal encroachment of land by other people who couldn't be removed straight away. Ad hoc reactive relocation also resulted in families being moved to lands that were later discovered to be prone to other hazards, such as flooding, landslides and cyclones.

Another major downfall of relocation in Sri Lanka was the negative impact on people's livelihoods. The reconstruction strategy was centred on "providing a house for a house", and livelihood aspects were not considered. Coastal communities who previously depended on the sea for their livelihoods, such as fishing and tourism, were adversely affected by being moved inland and away from the sea.

However, lessons learnt from the tsunami experience have led to improved land-use planning practices in Sri Lanka, with a good example being the Hambantota City Redevelopment Project. The Urban Development Authority in charge of land-use planning took the opportunity after the tsunami to review existing plans and introduced a new zoning system in the city of Hambantota based on hazard assessments. "No-development zones" were identified in high-risk areas, where alternate activities have been introduced for these areas, along with low-risk "safe areas" identified for new settlements.

It was a common phenomenon that people did not want to relocate due to their attachments with their lands and the community, and the impact it would have on their livelihoods. There was also confusion regarding new regulations implemented, which led people to ignore them to avoid inconveniences. The ambiguity of the buffer zone rule and the impact it had on people's livelihoods in Sri Lanka led some of the population to disregard regulations and illegally remain within the buffer zone.

The lack of recognition of potential hazards in pre-disaster times is a major cause of exacerbated impacts from hazard events. It is necessary to conduct multi-hazard assessments regularly and apply respective zoning regulations for land-use planning and exercise more caution in building in high-risk areas to protect communities.

The Sri Lankan example showed that applying post disaster reactive blanket reservations such as the coastal buffer zone are not effective for BBB. The main reason for the failure of the coastal buffer zone was the impracticality of this rule in terms of people's livelihoods, which led people to move back to coastal areas illegally. This case study illustrates the need for a comprehensive resettlement strategy which takes all these matters into consideration.

Case study: improving land-use planning in Christchurch after the 2010 and 2011 Canterbury Earthquakes

The first earthquake in the 2010 and 2011 Canterbury Earthquake sequence occurred on 4 September 2010. The earthquake had a magnitude of 7.1, but fortunately resulted in no fatalities. The first earthquake was followed by over 11,000 aftershocks and three major earthquakes on 26 December 2010, 22 February 2011 and 13 June 2011. Out of these, the 22 February 2011 quake was the most serious, resulting in 182 deaths, injuries and widespread damage to the built environment. The ground motions from the earthquakes led to significant soil liquefaction in east Christchurch and Kaiapoi. This resulted in extensive damage to homes and buildings built on these lands, roads, water infrastructure and other underground services.

The Canterbury Earthquake Recovery Authority (CERA) was established to plan, coordinate and implement the rebuild and recovery effort. One of the first tasks undertaken for the rebuild was a comprehensive geotechnical investigation of the land. As a result of the assessments, the entirety of Christchurch was divided into zones based on the soil conditions, exposure to seismic activity and probability of liquefaction. The recovery strategy included clear land-uses for each zone type:

- Red zone – Land with area-wide land and infrastructure damage that is too costly to repair, with a high probability of liquefaction. Red zoned areas were placed under a mandatory resettlement scheme, where homes, businesses and infrastructure were relocated out of all red zone areas.
- Green zone – Areas zoned green were considered suitable for reconstruction. Green zone lands were subdivided into four technical categories, each with specific technical guidelines provided to prevent future disruptions.

- Orange zone – Orange zoned lands required further investigations from engineers and were later re-classified as green or red.

The Department of Building and Housing published technical requirements for building in different land zones which were regulated through the building consent process. Updating land-use plans based on revised risk levels and coordinating them with building design regulations were excellent Build Back Better practices.

A difficulty experienced in the rebuild concerned the risk assessments having to be constantly updated due to ongoing aftershocks impacting soil stability, which caused delays and changes to the zoning process. However, there was good communication with relevant stakeholders, and planners and engineers were able to adapt. Flexibility and innovation among stakeholders were key.

The Christchurch rebuild opted to use highly skilled planners, engineers, designers and contractors for the rebuild. The adoption of a multi-stakeholder alliance approach to reconstruction, by creating the Stronger Christchurch Infrastructure Rebuild Team, facilitated frequent knowledge exchange and in-house training that assisted with overcoming challenges in the rebuild.

A Land Use Recovery Plan was developed by CERA to "provide direction for residential and business land use development to support recovery and rebuilding across metropolitan Christchurch for the next 10–15 years". This is a key long-term risk management system implemented following the earthquakes. The Land Use Recovery Plan focuses on relocating and developing new communities and suburbs in safer lower-risk areas over time. The plan states that Christchurch is to have space and infrastructure for 40,000 new houses by intensification of existing suburbs and proposing new or "greenfield" developments towards the west and inland areas of the city. Planning these future developments means that vital horizontal infrastructure, including water infrastructure, will need to be in place.

The Christchurch rebuild is a good example of successful cross-sector collaboration between engineering and planning for Building Back Better.

Land-use planning for Building Back Better

Land-use planning for Building Back Better requires consideration of a few different elements. The creation of risk zone maps based on accurate multi-hazard assessments which are coordinated with structural regulations is required for successful BBB. It is recommended that risk-based land-use plans and relevant building regulations are in place before reconstruction begins.

Avoiding high-risk lands for residential and commercial developments is desired if practical, and existing developments on high-risk lands should be encouraged to relocate to safer areas using appropriate strategies. "Buy-back schemes" and "land swap" schemes are possible options where high-risk lands can be bought by the government or swapped with lower-risk lands.

Past experiences have shown that relocation is a complicated and delicate process that can have adverse effects on a community's recovery if not handled properly. In terms of BBB, resettlement should be a last resort option, and needs to only be adopted alongside a comprehensive resettlement strategy. The resettlement strategy needs to take into account multi-hazard assessments and risk levels of new lands, community preferences, and include livelihood and lifestyle opportunities offered in the new locations. The resettlement strategy should be developed with community consent and input.

For new land-use plans to be effective, BBB advocates for the plans to be mandated using legislation and permit procedures. Educating the community and stakeholders about risk reduction, the recovery process, new regulations and providing support through workshops, training sessions and dissemination of multi-media information are fundamental for successful implementation and continued adoption of risk reduction practices.

Pre-disaster periods should also be used to implement BBB risk reduction measures to enhance the resilience of communities before a disaster strikes by identifying high-risk locations and incentivising existing developments to relocate. Ongoing multi-hazard assessments must be put in place, and these assessments should be used to create appropriate up-to-date planning and building regulations.

BBB Indicators for Multi-hazard-based Land-use Planning

Indicators or best practices for Building Back Better for Multi-hazard-based Land-use Planning were developed based on case study research findings and international examples. Land-use planning for BBB includes considerations for Risk-based Zoning and Resettlement. The BBB Indicators for land-use planning are listed in Table 4.1. The Indicators serve as a practical guide to direct

Table 4.1 Build Back Better Indicators for Multi-hazard-based Land-use Planning

Risk-based Zoning	Accurate, up-to-date understanding of all hazards in the area by conducting multi-hazard assessments and creating clear risk zone maps
	Land-use plans revised with appropriate land-uses determined by analysing the new risk zone maps and relevant updated building regulations, enforced using appropriate legislation
	Incentives such as land-swap schemes planned and put in place to facilitate and encourage the relocation of physical assets from high-risk to low-risk areas
	Education provided to stakeholders and the community on disaster risk and revised land-use plans prior to rebuilding
Resettlement	Create a comprehensive resettlement plan to minimise disruption and support the community through the resettlement process
	Consider the needs of the affected community
	Involve the community in choosing new sites
	Provide incentives for relocation (good infrastructure, employment opportunities etc.)
	Support the community through the provision of counselling and advisory services

stakeholders involved in post disaster activities to understand the elements that need to be considered when planning and implementing reconstruction and recovery programmes in order to Build Back Better.

References

2009 Victorian Bushfires Royal Commission (2010). *Final Report – Summary*. Melbourne, Australia: 2009 Victorian Bushfires Royal Commission.

Baradan, B. (2006). Analysis of the Post-disaster Reconstruction Process following the Turkish Earthquakes, 1999. In: Group, I. R. (ed.) *International Conference on Post-Disaster Reconstruction: Meeting Stakeholder Interests*. Montreal, Canada: University of Montreal.

Batteate, C. (2005) *International Symposium on Urban Disaster Risk Reduction and Regeneration Planning: Integrating Practice, Policy and Theory*. California Polytechnic State University, 3–5 November, San Luis Obispo, CA.

Bird, D. K., Chague-Goff, C. & Gero, A. (2011). Human Response to Extreme Events: A Review of Three Post-tsunami Disaster Case Studies. *Australian Geographer*, 42, 225–239.

Birkmann, J. & Fernando, N. (2008). Measuring Revealed and Emergent Vulnerabilities of Coastal Communities to Tsunami in Sri Lanka. *Disasters*, 32, 82–105.

Boano, C. (2009). Housing Anxiety and Multiple Geographies in Post-tsunami Sri Lanka. *Disasters*, 33, 762–785.

CERA (2011). *Draft Recovery Strategy for Greater Christchurch*. Christchurch, New Zealand: Canterbury Earthquake Recovery Authority.

De, L. L. (2011). *Post-disaster Reconstruction Strategy: Opportunity or Opportunism?* Masters of Science in Environmental Management, University of Auckland.

DFID (2004). *Disaster Risk Reduction: A Development Concern*. London: Department for International Development.

DMC, CCD & ADPC (2011). *Mainstreaming Disaster Risk Reduction into Approval Permits of Development Activities in the Coastal Areas of Sri Lanka*. Colombo, Sri Lanka: Disaster Management Centre, Coast Conservation Department and Asian Disaster Preparedness Centre.

DN & PA (2008). *Disaster and Development in South Asia: Connects and Disconnects. South Asia Disaster Report*. Colombo, Sri Lanka: Duryog Nivaran and Practical Action.

DPCD (2013). *List of Amendments to the Victoria Planning Provisions*. Melbourne, Australia: Victoria State Government Department of Planning and Community Development.

Frerks, G. & Klem, B. (2005). *Tsunami Response in Sri Lanka: Report on a Field Visit from 6–20 February 2005*. Wageningen, The Netherlands: Wageningen University and Clingdael University.

Glavovic, B. (2010). Realising the Potential of Land-use Planning to Reduce Hazard Risks in New Zealand. *Australasian Journal of Disaster and Trauma Studies*, 2010–1. Available: http://trauma.massey.ac.nz/issues/2010-1/glavovic.htm [Accessed 9 August 2018].

GoSL (2005). *Post-tsunami Recovery and Reconstruction Strategy*. Colombo, Sri Lanka: Government of Sri Lanka.

Guha-Sapir, D., Hargitt, D. & Hoyois, P. (2004). *Thirty Years of Natural Disasters – 1974–2003: The Numbers*. Louvain, Belgium: Presses Universitaires de Louvain.

Haigh, R., Amaratunga, D., Baldry, D., Pathirage, C. & Thurairajah, N. (2009). *ISLAND – Inspiring Sri Lankan Renewal and Development. RICS Research*. Salford, UK: University of Salford.

Halvorson, S. J. & Hamilton, J. P. (2010). In the Aftermath of the Qa'yamat: The Kashmir Earthquake Disaster in Northern Pakistan. *Disasters*, 34, 184–204.

Iglesias, G., Arambepola, N. M. S. I. & Rattakul, B. (2009). Mainstreaming Disaster Risk Reduction into Local Governance. In: *National Symposium on Creating Disaster Free Safer Environment*. Colombo, Sri Lanka: National Building Research Organisation and Ministry of Disaster Management and Human Rights.

Ikeda, S., Sato, T. & Fukuzono, T. (2007). Towards an Integrated Management Framework for Emerging Disaster Risks in Japan. *Natural Hazards*, 44, 267–280.

James Lee Witt Associates (2005). *Building Back Better and Safer: Private Sector Summit on Post-tsunami Reconstruction*, Washington, DC: James Lee Witt Associates.

Kates, R. W., Colten, C. E., Laska, S. & Leatherman, S. P. (2006). Reconstruction of New Orleans after Hurricane Katrina: A Research Perspective. *Proceedings of the National Academy of Sciences*, 103, 14,653–14,660.

Kennedy, J., Ashmore, J., Babister, E. & Kelman, I. (2008). The Meaning of "Build Back Better": Evidence from Post-tsunami Aceh and Sri Lanka. *Journal of Contingencies & Crisis Management*, 16, 24–36.

Mora, S. & Keipi, K. (2006). Disaster Risk Management in Development Projects: Models and Checklists. *Bulletin of Engineering Geology and the Environment*, 65, 155–165.

Morrow, B. H. (1999). Identifying and Mapping Community Vulnerability. *Disasters*, 23, 1–18.

Oliver-Smith, A. (1991). Successes and Failures in Post-disaster Resettlement. *Disasters*, 15, 12–23.

Olsen, S. B., Matuszeski, W., Padma, T. V. & Wickremeratne, H. J. M. (2005). Rebuilding after the Tsunami: Getting It Right. *AMBIO: A Journal of the Human Environment*, 34, 611–614.

Palliyaguru, R. & Amaratunga, D. (2008). Managing Disaster Risks through Quality Infrastructure and Vice Versa: Post-disaster Infrastructure Reconstruction Practices. *Structural Survey*, 26, 426–434.

Palliyaguru, R., Amaratunga, D. & Haigh, R. (2010). Integration of "Disaster Risk Reduction" into Infrastructure Reconstruction Sector: Policy vs Practise Gaps. *International Journal of Disaster Resilience in the Built Environment*, 1, 277–296.

Potangaroa, R. (2009). Native Engineering Technologies: The 2009 Samoan Tsunami and Its Significance for New Zealand. Unpublished report.

Quarantelli, E. L. (1987). Disaster Studies: An Analysis of the Social Historical Factors Affecting the Development of Research in the Area. *International Journal of Mass Emergencies and Disasters*, 5, 285–310.

Red Cross (2010). *World Disasters Report 2010 – Focus on Urban Risk*. Geneva, Switzerland: International Federation of Red Cross and Red Crescent Societies.

5 Early Warning and Disaster Risk Reduction Education

Introduction

Early Warning and Disaster Risk Reduction Education (DRRE) are aimed at bridging the gap between DRR knowledge and implementation in the community. Early Warning and DRRE have been highlighted many times in literature as key elements for Building Back Better and resilience.

Clinton's (2006) third proposition for BBB states that "Governments must enhance preparedness for future disasters" by putting in place robust systems that anticipate future disasters, and planning and preparedness mechanisms for future recovery processes. The *Hyogo Framework for Action 2005–2015* (HFA) (UNISDR, 2005) and the *National Disaster Recovery Framework* (FEMA, 2011) published by the United States Federal Emergency Management Agency attribute good governance, the use of risk knowledge to develop effective early warning systems, awareness-raising and education, changing practices and conditions that aggravate risk, and disaster preparedness through contingency plans, emergency funds and simulation exercises as resilience-building measures for communities (Shaw and Oikawa, 2014).

Building on the lessons learnt from implementing the HFA, the *Sendai Framework for Disaster Risk Reduction* (UNISDR, 2015) also advocates for early warning and DRR education through understanding and sharing disaster risk information, engaging appropriate stakeholders, and enhancing hazard early warning systems and preparedness.

Increasing community resilience can be achieved through raising risk reduction awareness and by implementing disaster preparedness mechanisms. Education and training on disasters and risk reduction enables communities to understand the importance of risk reduction measures as well as learn how to incorporate disaster capacity into their lives. Risk reduction training is essential for practitioners involved in rebuilding and recovery and development planning. Disaster preparedness plans allow people and businesses to be better prepared to respond to and recover from future disaster scenarios. Disaster preparedness mechanisms include the establishment of early warning systems, disaster management plans, risk-averse future development plans, and methods of alleviating climate change effects. Chapter 5 describes the different functions Early Warning and DRRE have in disaster risk reduction.

Early Warning

The United Nations *International Strategy for Disaster Reduction* defines early warning as "the provision of timely and effective information, through identified institutions, that allows individuals exposed to hazard to take action to avoid or reduce their risk and prepare for effective response" (UNISDR, 2004). Early warning systems are defined as a "set of capacities needed to generate and disseminate timely and meaningful warning information to enable individuals, communities and organisations threatened by a hazard to prepare and to act in sufficient time to reduce the possibility of harm or loss" (UNISDR, 2009). Early warning systems are designed to protect people by reducing risks and impacts as a result of actions taken before a disaster occurs (Basher, 2006).

Early warning requires technological infrastructure for data collection and analysis, decision support systems and decision analysis models, as well as social processes which involve decision-making (de León et al., 2006, Quansah et al., 2010, Horita et al., 2018). De León et al. (2006), ISDR-PPEW (International Secretariat for Disaster Reduction-Platform for the Promotion of Early Warning) (2005) and Basher (2006) propose four interrelated elements necessary for successful early warning:

- risk knowledge through systematic collection and analysis of data along with a dynamic assessment of hazards and physical, social, economic and environmental vulnerabilities;
- monitoring and warning based on good scientific methods for predicting and forecasting hazards with a warning system that operates continuously;

- communication and dissemination of warnings with clear messages and useful information to enable proper responses;
- response capability, ensuring that people understand their risks and how to react.

Having early warning systems in place alone is insufficient to achieve the desired reduction in disaster risk and impacts. Twigg (2003) states that the human factor in early warning systems is very significant, with the largest failures in early warning systems occurring due to issues with communication with and preparedness of local communities. Hurricane Katrina illustrated this with the failure to respect risk knowledge by the public as well as government organisations (Basher, 2006). Although meteorological warnings of the hurricane and potential storm surges were communicated in advance, the public's and government's response was inadequate due to an inability to understand the gravity and implications of the information received.

Early warning systems need to have strong political commitment, public support and institutional capacities. Early warning systems are ineffective if they are primarily top-down, hazard-focused, expert-driven systems with no end-user engagement (Basher, 2006). In order to be impactful, they need to capture the attention and interest of the public and need to be communicated in a manner that is easy to understand, inspiring action.

DRR Education

Disaster Risk Reduction Education attempts to link knowledge to action and "use knowledge, innovation and education to build a culture of safety and resilience at all levels", as highlighted in the *Hyogo Framework for Action*'s Priority 3 (UNISDR, 2005). Based on the lessons learnt from the HFA, the *Sendai Framework* declares that increasing public education and awareness of disaster risk is critical (UNISDR, 2015). The *Sendai Framework* also advocates the reduction of future disaster risk through Building Back Better.

In past disaster experiences, education and knowledge have provided local communities with the ability to reduce vulnerabilities and implement self-help strategies (Shaw et al., 2009). Children, community members and community leaders who have gained knowledge in disaster risk, mitigation and preparedness are better able to respond in a timely fashion to early warning messaging and circulate this information to others effectively.

The *Sendai Framework* proposes the adoption of DRRE through building the knowledge of government officials at all levels, promoting investments in innovation and technology development for DRRE and risk management, promoting the incorporation of disaster risk knowledge in formal and non-formal education, promoting national strategies and developing campaigns to strengthen public education and awareness, and investing in and developing people-centred disaster risk and emergency communications mechanisms and social technologies through a participatory process.

The *Sendai Framework* stresses that the consideration and participation of women, children and youth, persons with disabilities, indigenous peoples, migrants, older persons with valuable knowledge, skills and wisdom, academia and research entities, business, professional and private sector organisations, and the media are essential in creating effective policies and mechanisms for the development and dissemination of DRRE.

The challenge in DRRE is to successfully transfer knowledge to communities which will lead to action.

Illustrations of Building Back Better: successes and opportunities

Case study: tourism businesses in the Cook Islands

The regular cyclones that strike Rarotonga affect the tourism infrastructure and industry. As a result, tourism businesses have a lot of experience and knowledge of natural hazards and have innately developed a good understanding of what natural hazards are most likely to strike Rarotonga and the amount of time they have to prepare themselves following the first warnings. Businesses have also established direct contacts with the meteorological department through their community networks, and access first-hand information on cyclone tracking to keep themselves prepared.

The tourism businesses rely on experience and generational knowledge, and are aware and equipped for traditional preparatory practices such as placing shutters on windows and doors and tying roofs down in the event of a cyclone. Businesses like restaurants have also adopted adaptive and innovative practices such as using simple, open plan designs which allow wind and water to flow through, and using relocatable building components like toilets and bars that can be moved inland when cyclone warnings are issued.

Larger businesses such as hotels have formal evacuation and disaster response plans in place, which are supplemented by trained staff to assist evacuation procedures during an event. These businesses consider it their own responsibility to prepare these plans and perform regular drills to protect their clients and business.

The Tourism Industry Council in Rarotonga aids the disaster preparedness of businesses by having an accreditation system to assist with providing confidence and assurance of the safety of guests. However, there appears to be no formal risk reduction and preparedness information or training provided directly to businesses.

The Rarotonga case is a good example of the use of disaster risk reduction knowledge and adaptation for resilience. Although informal in this case, businesses use generational and first-hand disaster knowledge to prepare for and cope with disaster events. The Rarotongan tourism businesses have opted to rely on their own capacities, capabilities and knowledge to adopt various innovative and adaptive measures. These businesses could achieve even better outcomes if formal science-based information and training can be provided.

Case study: integration of seismic-resistant design and construction into the curricula of the Diploma in Associate Engineering (DAE) in Pakistan

The Kashmir Earthquake in 2005 and the 1935 Quetta Earthquake are listed in the top ten worldwide earthquakes in the last 100 years. Besides the significant death toll, the Kashmir event left 3.5 million people homeless due to the extensive damage to houses. A significant proportion of that vulnerability was borne by those in rural areas living in non-engineered houses, which were responsible for 75% of all earthquake fatalities. The 2005 earthquake made evident the need for inclusion of seismic design and construction in the engineering curricula to raise both technical standards and social awareness.

The intention to establish a new Diploma course for seismic-resistant design and construction was good BBB practice as part of DRRE, but there were challenges in developing and implementing it. There were different views on the inclusion of theory and practical content to target engineers as well as contractors, the types of materials covered (timber, concrete etc.), seismic retrofitting and seismic assessments of buildings, male–female student ratios and course costs which all had practical implications that needed to be considered thoroughly.

This example shows that the inclusion of BBB in educational programmes is not straightforward, but is a key aspect of long-term initiatives to resolve what can be deep-seated issues.

Case study: business recovery in Christchurch following the 2010/2011 Canterbury Earthquakes

Prior to the 2010/2011 earthquakes, there was no indication that earthquakes were a hazard in the Christchurch region. Therefore, most smaller businesses in Christchurch did not have any disaster preparedness plans in place when the earthquakes struck, and relied on standard insurance cover. Larger businesses that had health and safety plans in place found them useful for evacuation.

Following the experience of an ongoing sequence of earthquakes from 2010 to 2012, businesses began to develop their own preparedness plans. As a result, the preparedness plans of small businesses were mostly focused on evacuation and short-term issues, and did not consider any substantial risk reduction measures such as improving the resilience of their physical assets or considering moving to lower-risk areas. Overall, small businesses in Christchurch seem to have comparatively low awareness of early warning mechanisms and disaster risk reduction methods they can employ. Larger businesses seemed better informed, and addressed things like developing redundancy and not being over-reliant on technology, including adopting different communication and data-saving methods, as well as measures such as identifying alternative sites to be used as office spaces to enable businesses to operate even in the event of an emergency.

There have been council-led programmes to distribute information to the general public on relevant disaster-related issues and evacuation plans. However, more effective mechanisms to raise awareness of early warning systems and increase community preparedness for natural disasters are seen as necessary, targeting the business community in Christchurch. Public workshops or information evenings where businesses are invited to discuss issues around natural disasters, risks, evacuation and management plans could be effective.

Case study: school Emergency Management Plans in New Zealand

New Zealand is prone to several hazard types, with earthquakes and flooding being two of the most prominent ones. Therefore, the New Zealand Ministry of Education has developed templates for schools to create their own Emergency Management Plans for adoption during an emergency.

The Emergency Management Plans are designed to prepare teachers, staff and students at the schools for what to do in the event of an emergency.

Each Emergency Management Plan identifies:

- the school's evacuation plan based on type of event;
- an emergency contact plan for parents and caregivers;
- an external emergency contact list;
- checklists on response actions during and after emergency events such as fire, earthquakes, tsunamis, flooding, volcanic eruptions and ashfall, gas leaks, chemical spills and bomb threats;
- response actions for serious injury or death;
- response actions for missing students.

Getting schools and children involved in the development of these emergency management plans is a good DRRE mechanism as part of BBB.

Case study: early warning and DRRE initiatives for agricultural businesses in Gaza, Palestine

As part of a Build Back Better strategy for the rehabilitation of affected agricultural businesses in Gaza following the 2014 conflicts, it was identified that farmers should be engaged in the rebuilding and recovery process to raise awareness about the hazards and risks threatening them and to encourage them to start thinking differently to protect their agribusinesses from future risks.

Early warning and DRRE strategies proposed as part of BBB included:

- using local knowledge and traditional technologies for early warning;
- developing effective disaster risk and emergency communication methods and evacuation plans for future emergencies;
- introducing new/improved technologies to minimise the impact of disasters on businesses (i.e. physical assets and land-use);
- providing training to upgrade skills and knowledge of business-owners to use the new/improved technologies;

- training business-owners to make contingency plans on how to operate following a future disaster;
- educating on the importance of establishing strategic partnerships with other businesses to secure supply chains and collaborate to meet extra demands during emergency periods;
- introducing ways to back up and store important information and resources;
- introducing new capabilities to enable participation in reconstruction and recovery efforts (building skills etc.);
- providing training in BBB, DRR, disaster response and disaster recovery.

Early Warning and DRR Education for Building Back Better

Early Warning and DRR Education complements improving Structural Resilience and Multi-hazard-based Land-use Planning for disaster risk reduction as part of Building Back Better.

Where developments have to remain in high-risk lands, and if there is an inability to provide sufficient protection for the built environment, early warning mechanisms are indispensable to inform and direct people to safety.

Local communities as well as stakeholders involved in serving communities require DRR Education to gain a better understanding of realistic hazards and the risks to be expected, and learn how to respond to, withstand and recover from disaster events. DRR Education is also necessary to teach the importance of BBB initiatives for building resilience and to encourage adoption of BBB practices for pre-disaster and post disaster planning.

BBB Indicators for Early Warning and DRR Education

Indicators or best practices for Building Back Better for Early Warning and DRR Education were developed based on research findings and international examples. The BBB Indicators for Early Warning and DRR Education are listed in Table 5.1. The Indicators serve as a practical guide to direct stakeholders

Table 5.1 Build Back Better Indicators for Early Warning and DRR Education

Early Warning	Advanced science and local knowledge are used to improve the accuracy of hazard predictions
	Early warning systems appropriate to the local community are employed using a combination of new technology and local knowledge and traditional methods
DRR Education	Communities and stakeholders are educated on prevalent disaster risk, practical disaster risk reduction methods and disaster preparedness
	Training and resources provided to organisations on preparing business continuity plans and establishing partnerships with other organisations for improved resilience

involved in post disaster activities to understand the elements that need to be considered when planning and implementing reconstruction and recovery programmes in order to Build Back Better

References

Basher, R. (2006). Global Early Warning Systems for Natural Hazards: Systematic and People-centred. *Philosophical Transactions of the Royal Society of London A: Mathematical, Physical and Engineering Sciences*, 364, 2,167–2,182.

Clinton, W. J. (2006). *Lessons Learned from Tsunami Recovery: Key Propositions for Building Back Better*. New York: Office of the UN Secretary-General's Special Envoy for Tsunami Recovery.

de León, J. C. V., Bogardi, J., Dannenmann, S. & Basher, R. (2006). Early Warning Systems in the Context of Disaster Risk Management. *Entwicklung und Ländlicher Raum*, 2, 23–25.

FEMA (2011). *National Disaster Recovery Framework: Strengthening Disaster Recovery for the Nation*. Washington, DC: Federal Emergency Management Agency.

Horita, F. E. A., de Albuquerque, J. P. & Marchezini, V. (2018). Understanding the Decision-making Process in Disaster Risk Monitoring and Early-warning: A Case Study within a Control Room in Brazil. *International Journal of Disaster Risk Reduction*, 28, 22–31.

ISDR-PPEW (2005). *The International Early Warning Programme – IEWP*. Bonn, Germany: UN/International Secretariat for Disaster Reduction Platform for the Promotion of Early Warning. Available: https://www.unisdr.org/2006/ppew/info-resources/docs/IEWP.pdf [Accessed 9 August 2018].

Quansah, J. E., Engel, B. & Rochon, G. L. (2010). Early Warning Systems: A Review. *Journal of Terrestrial Observation*, 2, 24–44.

Shaw, R. & Oikawa, Y. (2014). *Education for Sustainable Development and Disaster Risk Reduction*. Tokyo, Japan: Springer.

Shaw, R., Takeuchi, Y. & Rouhban, B. (2009). Education, Capacity Building and Public Awareness for Disaster Reduction. In: Sassa, K. & Canuti, P. (eds) *Landslides – Disaster Risk Reduction*. Berlin, Germany: Springer.

Twigg, J. (2003). The Human Factor in Early Warnings: Risk Perception and Appropriate Communications. In: Zschau, J. & Kuppers, A. (eds) *Early Warning Systems for Natural Disaster Reduction*. Berlin, Germany: Springer.

UNISDR (2004). *International Strategy for Disaster Reduction*. Geneva, Switzerland: United Nations Inter-Agency Secretariat for the International Strategy for Disaster Reduction.

UNISDR (2005). *Hyogo Framework for Action 2005–2015: Building the Resilience of Nations and Communities to Disasters*. Kobe, Japan: United Nations Office for Disaster Risk Reduction.

UNISDR (2009). *2009 UNISDR Terminology on Disaster Risk Reduction*. Geneva, Switzerland: United Nations Inter-Agency Secretariat for the International Strategy for Disaster Reduction.

UNISDR (2015). *Sendai Framework for Disaster Risk Reduction 2015–2030*. Sendai, Japan: United Nations Office for Disaster Risk Reduction.

6 Psychological and Social Recovery

Introduction

This chapter looks at the importance of supporting the psychological and social recovery of people for overall disaster recovery and BBB. It illustrates mechanisms for supporting and empowering affected people and their positive impacts on recovery and resilience. The destruction faced by communities from natural and man-made disasters is multi-faceted. The most noticeable damage is to the built environment and economy. The impact of disasters on a community's psychological state and social life is less visible, yet has a significant effect on overall recovery (Kristensen, 2012, Cook et al., 2008). Affected people suffer from post-traumatic stress disorder and prolonged grief disorder which affect their decision-making abilities and motivation to move forward with recovery (Kristensen, 2012).

Common post disaster social issues and the need to BBB

Post disaster recovery focuses on providing fast solutions in an attempt to re-establish a sense of normality in affected communities as soon as possible (Khasalamwa, 2009). The focus on speed can result in overlooking the real needs of communities (Baradan, 2006, de Silva, 2009). The community are often not consulted to provide their input on reconstruction and recovery (Waugh and Smith, 2006, Boano, 2009). The lack of community consultation and participation leads to the provision of recovery solutions that are not suitable (Boano, 2009, Waugh and Smith, 2006). For example, some of the new houses constructed in Sri Lanka by humanitarian agencies during the Indian Ocean Tsunami rebuild featured bathrooms made with half-height walls and shared bathrooms for males and females which were culturally unacceptable (Ruwanpura, 2009). Locals were unhappy with the reconstruction of homes following the 1999 Marmara Earthquake in Turkey as their local life, culture and aesthetics were not considered (Tas, 2010).

Non-participatory resettlement for risk reduction also creates substantial issues (Oliver-Smith, 1991). In Sri Lanka and Samoa following the respective tsunami disasters, resettlement of coastal communities inland has led to the loss of traditional livelihoods such as fishing and tourism which has impacted families and their ability to recover (Frerks and Klem, 2005, Potangaroa, 2009). Inequalities in the provision of aid and support to different ethnic groups during resettlement in Sri Lanka created tensions and conflicts (Ruwanpura, 2009). Colten et al. (2008) and Khasalamwa (2009) state that insufficient attention to social, cultural and ethnic facets of communities during recovery exacerbate pre-existing vulnerabilities.

The trauma faced by disaster victims needs to be recognised as a serious issue during recovery (Matanle, 2011). In most developing countries where formal psychological support is not common, community networks provide empathetic support to each other (Asian Development Bank et al., 2005). Disasters and resettlement operations can impact upon psychological recovery if the community are separated by disrupting community cohesion (Florian, 2007). The involvement of communities in recovery, such as through owner-building, is used as a mechanism to support psychological recovery (Kennedy et al., 2008). However owner-building without training and supervision has created sub-standard homes during the tsunami rebuild in Sri Lanka (Pathiraja and Tombesi, 2009). External tradespeople with construction expertise employed for rebuilding to avoid such problems also leave the locals excluded and disconnected from feeling a sense of responsibility about the community's recovery (Oliver-Smith, 1991), indicating that community involvement to a certain degree is necessary for successful recovery.

The strategies that community members tend to adopt for Building Back Better depend on the resources they can access, their expectations about their community's prospects for recovery, and ultimately, community-level collective narratives (Chamlee-Wright and Storr, 2011). Aldrich examined the post disaster responses of four distinct communities – Tokyo following the 1923 earthquake, Kobe after the 1995 earthquake, Tamil Nadu after the 2004 Indian

Ocean Tsunami, and New Orleans post-Katrina – and found that those with robust social networks were better able to coordinate recovery (Aldrich, 2012).

In disasters, family ties are central to resilience because family members commonly serve as the first providers of assistance (Drabek and Boggs, 1968, Garrison and Sasser, 2009, Beggs et al., 1996). Individuals assume that family members, especially immediate family members, will support each other in disasters, 36% identifying only family members among their social capital networks for disaster assistance in a study conducted by Meyer (2013). Disaster scholars have used social capital to understand the trajectory of individuals (based on what resources are accessed through social networks) as well as communities (based on levels of trust, collective action and other public goods). Social networks provide financial (e.g. loans and gifts for property repair) and non-financial (e.g. search and rescue, debris removal, child care during recovery, emotional support, sheltering and information) resources (Aldrich and Meyer, 2015). The social bonding serves as social insurance. Bonding social capital can reduce individuals' likelihood of seeking formal aid from organisations during disasters (Beggs et al., 1996). The community are put at the heart of the BBB Framework by Mannakkara and Wilkinson, where psycho-social elements of community are at the core of recovery.

Illustrations of Building Back Better: successes and opportunities

Case study: community recovery in Fiji following Tropical Cyclone Winston, 2016

Tropical Cyclone Winston hit Fiji on 20 and 21 February, 2016. The Category 5 cyclone resulted in 44 deaths and the evacuation of over 62,000 people to nearly 900 evacuation centres. The cyclone damaged or destroyed an estimated 28,000 houses as it passed through the Fiji Islands. The discussions among the affected towns and villages once they had stabilised focused on how to get back to where they were and how they could find and activate a resilient response.

Rakiraki was the Fijian district at ground zero of the affected area on the main island of Viti Levu. The Fiji Red Cross Society (FRCS) sub-office in Rakiraki had been working in several villages, and it was suggested that a village in Nokonoko could be used to study recovery resilience. The village had been heavily damaged during the cyclone, with over 60% of houses losing their roofs.

The village was surveyed for the project in June 2016. Health and social programmes were being run by the FRCS together with a strong and obvious community network based on family and kinship. There was ongoing information on programmes such as the government's "Help for Homes", which was widely known within the village. In addition, ten Red Cross volunteers lived in the village, so it was well connected in terms of information. There was strong participatory decision-making through the village Council of Leaders, many had already started working on the repair of their houses, and the village school was re-opened by June 2016. The chief of the village was also the chairperson of the FRCS Rakiraki sub-office, which meant that the village had strong connections with the FRCS, which assisted the recovery process.

There appeared to be a strong social connectedness in the village, and the expectation was that it was "Building Back Better" as a result. The DASS42 Quality of Life Survey was conducted with 88 people from the village to test the psychosocial recovery aspects of the villagers. The Depression Anxiety Stress Survey (DASS42) is a tool developed by Lovibond at the University of New South Wales, Australia which has been used extensively to examine and support community recovery initiatives in many countries. The DASS42 survey provides information about the levels of depression, anxiety and stress experienced by people.

The results from the DASS42 survey conducted in the village in Nokonoko were:

- Depression – 15.0 Moderate
- Anxiety – 14.9 Severe
- Stress – 18.4 Mild

The elevated depression and high anxiety scores indicated that there could be issues with regard to recovery despite the apparent positive indicators such as rebuilding progress, community involvement and strong community networks. This result was informally confirmed by the ten Red Cross staff who lived in the village. The survey did not highlight any differences between the impact on men or women, nor did it highlight differences in the ages of the people in the village in terms of their psychological and social recovery.

Sixty-six per cent of villagers reported that they had good family connections and partook in volunteer work; however, 34% of villagers reported that they were alone or lonely for considerable periods, with 15% stating that they were alone or lonely all of the time. This apparent loneliness or social disconnection was directly linked to a significantly lower Quality of Life. This suggested that putting in place specific initiatives to support psychological and social recovery plays a major role in Building Back Better.

Villagers selected housing, health, schooling and education as the key issues for their families, where "schooling" referred to attendance at the local school, and "education" was attending school outside the village. From the people's perspective, rebuilding homes, getting schools operating and bringing community-based health programmes into the villages were recovery priorities.

It is important to understand the local context and conditions when planning for recovery in order to deliver recovery solutions that target the community's vulnerabilities and encourage building resilience. For example, in the village studied, the average family size was 5.9 and the number of people per bedroom was 3.4, suggesting a level of overcrowding. This could or would be reflected in the health statistics. The average household income reported was 0–5,000 Fijian dollars per year, and a basic 6×4-metre shelter cost in the order of 8,500–10,000 Fijian dollars. The unaffordability of rebuilding was the reason why shelter was their number one priority. The low income also explains why 60% of the village relied entirely on traditional food sources and what could be grown and gathered, and 74% relied on these for a considerable period of the time or more.

Understanding local conditions also presents the community's strengths. For example, in this village, 34% reported that the highest educational qualification in their family was university-based, 43% reported technically-based or higher, and 88% secondary or higher. This suggests that there were skills that could be developed from within the village to aid the recovery process.

There are ostensible underlying issues that the village will need to address on its Build Back Better "journey", and the BBB Framework (or any framework) needs to be seen in relation to what is happening on the ground.

Case study: post disaster social recovery following the 2009 Victorian Bushfires

The recovery model for Victoria centred on the community. The guiding principles for the recovery effort consisted of: safety and welfare, meeting needs, community engagement, fairness and equity, and tailored solutions to suit local communities.

One of the first lines of support provided to bushfire-affected residents was the launch of a case management service initiated by the Victorian Bushfire Reconstruction and Recovery Authority (VBRRA). Each affected family was assigned a case manager who provided information and direction for recovery. Victims of the bushfires that were assisted by the case management service felt it was one of the most important forms of support they received in the recovery process. The bushfire experience showed that case managers were very valuable in supporting psycho-social recovery of people and should be a part of recovery efforts. The use of trained case managers who are familiar with the local community context and having the service operate for longer through long-term recovery were recommended from lessons learnt.

Along with the case management service, there were many initiatives that focused on supporting psychological and social recovery in Victoria. The Victorian Bushfire Appeal Fund (VBAF) was arranged by VBRRA to collect donations and fund community projects such as community halls, recreational facilities, schools and medical centres. Community hubs (information centres) were set up to provide direction for advice, financial support, counselling services, and to facilitate social interaction between community members. The Rebuilding Advisory Service (RAS) was established to provide rebuilding advice, and business information support centres and business mentoring services to assist business-owners. Information was also provided through newsletters, other media and training courses to educate people on building requirements and bushfire risk.

Social recovery also included organising commemorative events and memorials as desired by the affected communities. The one-year anniversary memorial was a major turning point, and it was observed that people began to move forward with their lives and started making decisions following this event. The speedy construction and opening of public buildings was a positive step towards rebuilding confidence in residents to remain in these communities and rebuild their lives again. Activities such as retreats and workshops to support individual groups such as men, women, youth, families and schoolchildren were also held, to promote social recovery. Temporary villages which housed groups of people were one form of temporary accommodation provided, which aided psycho-social recovery because they helped to create a sense of community.

Providing counselling services through the community hubs was an important intervention. Similar to other countries, following the bushfires people were grieving and unable to make decisions regarding their future, which slowed down overall recovery progress in terms of housing reconstruction and business rebuilding. Therefore, recognising and supporting the community's psychological needs has a big impact on recovery progress and success.

Along with supporting the community, the importance of involving the community in the recovery process was understood. The recovery in Victoria was designed to be community-centred, with community consultation and engagement used to provide tailor-made recovery solutions. The first step taken to enable community involvement included a series of community meetings held at the start of the recovery process to identify the specific rebuilding and recovery needs of each affected community. The meetings allowed the community to discuss matters such as each town's identity, what should be retained, and future visions for the town. This information went back to the VBRRA, after which recovery projects were decided upon based on funds.

The creation of community recovery committees (CRCs) consisting of nominated elected members of the community was another tool used for community involvement. The CRCs worked through Community Recovery Plans to sort out projects and ideas with feedback received from and distributed to the wider community through the CRC members. This feedback was again used by the VBRRA to develop recovery and rebuild projects. This type of involvement and consultation can be a tricky process if there are pre-existing community divisions that can lead to disagreements. Therefore, maintaining impartiality, wide community involvement and strong decision-making by the local council or recovery authority are required.

Another strategy employed for community involvement was the "owner-building" approach taken towards housing reconstruction. It was the responsibility of home-owners to employ builders (volume builders or small-scale builders) for housing reconstruction. The home-owners acted as project managers and were responsible for managing funds (insurance and VBAF grants) and the builders and subcontractors used for the rebuild. This process created many issues for home-owners who were inexperienced in managing a build and working with builders and contractors. The RAS was therefore established to support home-owners and provide advice and guidance on the rebuilding process, and became a very useful BBB initiative.

Most communities are very attached to the particular identities of their cities, towns and villages. This was the case in Marysville, one of the badly affected towns, whose inhabitants were attached to their identity as a quaint mountainous village and did not want to see it change. The rebuild took the opportunity to upgrade and modernise local buildings and infrastructure, but the residents of Marysville were unhappy with the use of modern architecture for their important public buildings, such as the school and police station. The local council is also unable to afford long-term maintenance of the modern infrastructure put in place. This example reiterates the importance of community and local council involvement in recovery planning from the beginning, to avoid such problems.

On the other hand, the problem with enabling a high level of community input and control is that disaster victims are sometimes not capable of decision-making due to grief and trauma. The case managers, builders and RAS rebuilding advisors who worked in the bushfire-affected communities observed that people were not emotionally ready to do anything. The decision-making and process around rebuilding their homes put a large amount of pressure on locals. Many people remained in temporary accommodation for more than a year, and there were many half-built homes observed in the two- to three-year period following the bushfires.

The bushfires experience displayed many good initiatives to support psychological and social recovery for BBB. The key lessons learnt show that community involvement needs to be encouraged, along with sufficient community support, to achieve best results. Understanding the local context and local community dynamics are also important in recovery planning for BBB.

Case study: improving community wellbeing after the 2015 Nepal earthquakes

The 2015 Nepal Earthquake occurred on 25 April, with its epicentre at Barpak, Gorkha, and a major aftershock on 12 May. The earthquake killed nearly 9,000 people and injured 22,000. The earthquake also triggered an avalanche on Mount Everest and in the Langtang valley, resulting in further deaths. The earthquake displaced 3.5 million people and destroyed UNESCO World Heritage Sites. The total damage from the earthquake was estimated to be US$10 billion, which was 50% of the country's GDP.

Active community participation was observed after the earthquakes with volunteers from the community involved in the immediate rescue. People were resourceful and used locally available materials to construct temporary shelters. The community were also active in preparing food and distributing water. Traditionally, Nepalese communities have strong networks with family and neighbours. Identity, nationalism, friendship and neighbourhood are factors of social bonding. Inter-community networks incorporate various social relationships such as marriages and festivals, forming a bridge between different communities. In Nepalese communities, there is a strong relationship formed after marriages between members of different communities, which brings different communities together. Bridging activities such as festivals and associations bring together individuals from different locations, identities and language groups. The communities of Nepal are composed of different ethnicities and castes, but they share the same culture and religious and social values in many cases. Cities are often segregated into heterogeneous occupation classes or caste groups; however, there is social harmony among inhabitants. The importance of social relationships, trust, respect and value for each other enhanced the recovery process.

Building better community cohesiveness and assisting community-based recovery helps build resilience to disasters by providing the community with the ability to withstand and overcome the impacts of disaster events. Social cohesion, social bonds, relationships, communications and strong neighbourhoods are important for enhanced recovery.

In terms of community support for BBB, affected families received financial assistance either from the government or from NGOs. As recommended by the steering committee of the National Reconstruction Authority (NRA), the housing grant for families who lost their homes during the earthquakes were increased to US$3,000 from the previous amount of US$2,000. The NRA began distributing the first instalment of the housing grant in March 2016, distributing grants to 407,004 families. One of the strategic recovery objectives in the *Post-Disaster Recovery Framework* (PDRF) published by the NRA (2016) was to "Strengthen the capacity of people and communities to reduce their risk and vulnerability, and to enhance social cohesion". The framework addressed the implementation of effective mainstreaming of gender equity, social inclusion, closing the gender gap, participation of women and marginalised groups, and to raise awareness and the capacity of women and vulnerable and marginalised groups. There was no specific strategic framework from the government and NRA to provide support and counselling for mental health, but some indigenous and international NGOs had been working to provide psychological support and mental health services prior to the disaster.

One of the focuses of the PDRF was social cohesion. There was an exponential increment in social cohesion and strength immediately after and during the disaster period, but this began to decline following the early post disaster period. Therefore, maintaining this social cohesion through external interventions is important even for well-connected communities, and is indispensable for rebuilding community cohesion and connectivity for Building Back Better.

In terms of community involvement, the disaster recovery practice in Nepal is supply-driven rather than participatory, and follows a top-down approach. Therefore, there was almost non-existent community participation in the formulation of disaster recovery plans and the recovery process. The community members were unaware of the plans and strategies developed for the recovery. If the community were included in the formulation and implementation of recovery plans, better recovery outcomes could have been achieved. There was also no local government involvement with recovery, as government authorities, including the reconstruction authority, lacked locally elected members.

The immediate post disaster response was successful in Nepal as a result of the high level of social cohesion. However, long-term recovery has been slow and inefficient. To Build Back Better and create recovery outcomes that are favourable to the community in the long term, it is necessary to involve and include the local community to some degree in recovery planning, as well as support the community with their recovery process.

Case study: social capital for the tourism sector recovery and resilience in the Cook Islands

The recovery of tourism businesses in Rarotonga following the frequent coastal disasters experienced in the Cook Islands is crucial to the recovery of the country as a whole. Social resources such as community networks and personal and professional contacts have a significant positive impact on post disaster recovery. These social resources are present in exceptional quality and volume in the Cook Islands. The country has strong societies and strong village networks. During and after disaster events, the local community are the first responders and a collective effort is made to check on the community, pool resources and provide assistance as necessary. This includes businesses supporting each other. For example, hotels would take in guests from neighbouring hotels if they have been impacted to a greater extent and are not coping.

Therefore, a high level of social cohesion is a major factor in BBB and resilient post disaster recovery.

Psychological and Social Recovery for Building Back Better

There is a new paradigm for disaster recovery ahead of traditional intervention on recovery tools of brick, mortar and concrete. Social issues arising in post disaster environments related to social, cultural, religious, ethnic and psychological factors impact the success of recovery. These issues need to be addressed in the recovery process for BBB by supporting as well as empowering the community by involving them in recovery.

Reconstruction is a chaotic and stressful time for individuals who are also experiencing trauma. These communities require various forms of assistance, such as personalised advice and support, special support for vulnerable communities, psychological support and counselling. Keeping the community informed is also necessary to aid community recovery. Community cohesion and social capital are key contributors to resilience, therefore recovery in line with BBB needs to incorporate activities to build cohesion within communities.

Each community is unique, therefore one of the first steps to be taken in post disaster recovery planning is to understand the local context of the affected community through needs assessments and surveys in order to provide appropriate assistance tailored to each community. Reconstruction and recovery policies must then be developed based on local requirements to support and preserve the local culture and heritage. Maintaining community involvement throughout recovery is integral to success. The level of community involvement can vary depending on the recovery approach taken and the culture and requirements of the community. Highly decentralised recovery efforts are typically understood to be more in line with BBB as they place the community at the centre of recovery and they aim to empower the community by supporting them to set up and execute their own projects. At the other end of the spectrum are highly centralised recovery efforts, led by the government and NGOs, which initiate and execute recovery projects with little to no input or involvement from the community. In this type of recovery, the community should at least be kept informed and educated about the recovery process through regular workshops. Advantages and disadvantages have been observed in adopting both highly decentralised efforts and highly centralised efforts depending on the community and its preconditions, and often a combination of the two tailored to suit the community is most appropriate.

70 *Psychological and Social Recovery*

In any case, community consultation should be an essential component of recovery planning, and the creation of community consultation groups consisting of community leaders from pre-existing community groups and reputed members of the community to liaise between the wider community and governmental authorities can be an option. Existing community groups can also be called upon to assist with recovery activities. The government should maintain full transparency with the affected communities about recovery plans, issues and solutions so that the communities have full awareness and are able to make educated decisions. Although owner-building has been recognised as problematic (Pathiraja and Tombesi, 2009), authors such as Lloyd-Jones (2006), Olshansky (2005) and Ozcevik et al. (2009) propose that if it is supported with proper training, thorough supervision and advice, owner-building could be a good way to include the community in the recovery process.

BBB Indicators for Psychological and Social Recovery

Indicators or best practices for Building Back Better for Psychological and Social Recovery were developed based on case study research findings and international examples. Psychological and Social Recovery for BBB takes the form of Community Support and Community Involvement to build community resilience. The BBB Indicators for Psychological and Social Recovery are listed in

Table 6.1 Build Back Better Indicators for Psychological and Social Recovery

Community Support	Establish community advisory services to provide information and connect with affected households
	Identify vulnerable groups in the community and organise specialised assistance to support them
	Organise psychological support and counselling services for the community
	Organise activities and support groups to bring the community together and build social cohesion. Create a sense of community and togetherness
	Inform the community regularly on recovery decisions and progress using appropriate channels (e.g. regular public meetings, pamphlets, newsletters, media, text messages, dedicated recovery website, social media)
	Prioritise the rebuilding of public facilities and heritage sites based on community's social and cultural needs (e.g. schools, churches, hospital, supermarket, community halls, recreation centres)
Community Involvement	Empower the community by incorporating grass-roots methods for recovery (e.g. creating/utilising community groups to get community input for planning, decision-making and various aspects of implementation of rebuild and recovery)
	Promote owner-building of homes to empower home-owners with support and supervision from skilled builders
	Maintain full transparency with affected communities with regard to recovery decisions

Table 6.1. The Indicators serve as a practical guide to direct stakeholders involved in post disaster activities to understand the elements that need to be considered when planning and implementing reconstruction and recovery programmes in order to Build Back Better.

References

Aldrich, D. P. (2012). *Building Resilience: Social Capital in Post-disaster Recovery*. Chicago, IL: University of Chicago Press.

Aldrich, D. P. & Meyer, M. A. (2015). Social Capital and Community Resilience. *American Behavioral Scientist*, 59, 254–269.

Asian Development Bank, Japan Bank for International Cooperation & World Bank (2005). Preliminary Damage and Needs Assessment. *Sri Lanka 2005 Post-Tsunami Recovery Programme*. Colombo, Sri Lanka: Asian Development Bank, Japan Bank for International Cooperation and World Bank.

Baradan, B. (2006). Analysis of the Post-disaster Reconstruction Process following the Turkish Earthquakes, 1999. In: Group, I. R. (ed.) *International Conference on Post-Disaster Reconstruction: Meeting Stakeholder Interests*. Montreal, Canada: University of Montreal.

Beggs, J. H., Haines, V. A. & Hurlbert, J. S. (1996). Situational Contingencies Surrounding the Receipt of Informal Support. *Social Forces*, 75, 201–222.

Boano, C. (2009). Housing Anxiety and Multiple Geographies in Post-tsunami Sri Lanka. *Disasters*, 33, 762–785.

Chamlee-Wright, E. & Storr, V. H. (2011). *Social Capital as Collective Narratives and Post-disaster Community Recovery*. Oxford, UK: Blackwell.

Colten, C. E., Kates, R. W. & Laska, S. B. (2008). Three Years after Katrina: Lessons for Community Resilience. *Environment*, 50, 36–47.

Cook, A., Watson, J., Van Buynder, P., Robertson, A. & Weinstein, P. (2008). Natural Disasters and Their Long-term Impacts on the Health of Communities. *Journal of Environmental Monitoring*, 10, 167–175.

de Silva, M. W. A. (2009). Ethnicity, Politics and Inequality: Post-tsunami Humanitarian Aid Delivery in Ampara District, Sri Lanka. *Disasters*, 33, 253–273.

Drabek, T. E. & Boggs, K. S. (1968). Families in Disaster: Reactions and Relatives. *Journal of Marriage and Family*, 30, 443–451.

Florian, S. (2007). Housing Reconstruction and Rehabilitation in Aceh and Nias, Indonesia – Rebuilding Lives. *Habitat International*, 31, 150–166.

Frerks, G. & Klem, B. (2005). *Tsunami Response in Sri Lanka: Report on a Field Visit from 6–20 February 2005*. Wageningen, The Netherlands: Wageningen University and Clingdael University.

Garrison, M. E. B. & Sasser, D. D. (2009). Lifespan Perspectives on Natural Disasters: Coping with Katrina, Rita, and Other Storms. In: *Families and Disasters: Making Meaning Out of Adversity*. New York: Springer.

Kennedy, J., Ashmore, J., Babister, E. & Kelman, I. (2008). The Meaning of "Build Back Better": Evidence from Post-tsunami Aceh and Sri Lanka. *Journal of Contingencies & Crisis Management*, 16, 24–36.

Khasalamwa, S. (2009). Is "Build Back Better" a Response to Vulnerability? Analysis of the Post-tsunami Humanitarian Interventions in Sri Lanka. *Norwegian Journal of Geography*, 63, 73–88.

Kristensen, P. L. T. (2012). Bereavement and Mental Health after Sudden and Violent Losses: A Review. *Psychiatry: Interpersonal & Biological Processes*, 75, 76–97.

Lloyd-Jones, T. (2006). *Mind the Gap! Post-disaster Reconstruction and the Transition from Humanitarian Relief*. London: Royal Institution of Chartered Surveyors.

Matanle, P. (2011). The Great East Japan Earthquake, Tsunami, and Nuclear Meltdown: Towards the (Re)construction of a Safe, Sustainable, and Compassionate Society in Japan's Shrinking Regions. *Local Environment*, 16, 823–847.

Meyer, M. A. (2013). *Social Capital and Collective Efficacy for Disaster Resilience: Connecting Individuals with Communities and Vulnerability with Resilience in Hurricane-prone Communities in Florida*. Doctor of Philosophy, Colorado State University.

NRA (2016). *Post-Disaster Recovery Framework*. Kathmandu, Nepal: Nepal Government.

Oliver-Smith, A. (1991). Successes and Failures in Post-disaster Resettlement. *Disasters*, 15, 12–23.

Olshansky, R. B. (2005). How Do Communities Recover from Disaster? A Review of Current Knowledge and an Agenda for Future Research. *46th Annual Conference of the Association of Collegiate Schools of Planning*. Kansas City, MO.

Ozcevik, O., Turk, S., Tas, E., Yaman, H. & Beygo, C. (2009). Flagship Regeneration Project as a Tool for Post-disaster Recovery Planning: The Zeytinburnu Case. *Disasters*, 33, 180–202.

Pathiraja, M. & Tombesi, P. (2009). Towards a More "Robust" Technology? Capacity Building in Post-tsunami Sri Lanka. *Disaster Prevention and Management*, 18, 55–65.

Potangaroa, R. (2009). Native Engineering Technologies: The 2009 Samoan Tsunami and Its Significance for New Zealand. Unpublished report.

Ruwanpura, K. N. (2009). Putting Houses in Place: Rebuilding Communities in Post-tsunami Sri Lanka. *Disasters*, 33, 436–456.

Tas, M. (2010). Study on Permanent Housing Production after 1999 Earthquake in Kocaeli (Turkey). *Disaster Prevention and Management*, 19, 6–19.

Waugh, W. L. & Smith, R. B. (2006). Economic Development and Reconstruction on the Gulf after Katrina. *Economic Development Quarterly*, 20, 211–218.

7 Economic Recovery

Introduction

Disasters cause damage to the economy of communities, with the disruption of businesses and income-generating industries leading to issues such as high inflation rates and poverty. The adverse effects of disasters on the economy can also impede the overall recovery of a city. Economic recovery is often outshone by the attention given to built environment reconstruction. Building Back Better is often interpreted as a concept that focuses on improving the structural resilience of buildings and infrastructure, but authors such as Lewis (2002), de Silva (2009), Kennedy et al. (2008) and Mannakkara and Wilkinson (2012) stress the importance of supporting economic recovery and rejuvenation of communities as an important part of BBB. Economic recovery of a town or city following a disaster affects the recovery decisions made by members of the community. Often, the rate of economic recovery influences the rebuilding of homes and infrastructure and the community's choices regarding relocation.

74 *Economic Recovery*

This chapter presents how economic recovery can be best supported in local affected communities for BBB and achieving community resilience. It explores the development of a resilient economic recovery strategy which considers effective funding mechanisms, support for local businesses, and protection and enhancement of livelihoods. BBB propositions for economic recovery, which serve as practical recommendations for implementation developed through international case study research and literature review are presented, along with case study examples demonstrating these propositions in action.

Common issues with post disaster economic recovery

Post disaster recovery efforts to date have shown support for economic recovery with strategies such as cash-for-work programmes, provision of business grants, asset replacement programmes to provide industries with necessary resources, and training programmes to up-skill locals and help them find work (de Silva, 2009, VBRRA, 2010, Haigh et al., 2009). In Aceh, Indonesia, tsunami-affected people were trained and employed in reconstruction to provide them with a source of income alongside the opportunity to become involved in their own recovery (Kennedy et al., 2008). In Japan following the 2011 Earthquake and Tsunami, the government decided to consolidate smaller fishing markets into large fishing centres to enable fishermen to support each other (Okuda et al., 2011). Christchurch City Council's *Central City Plan* in New Zealand proposed fast-tracking of building consents for businesses to allow quicker repair and construction work (Christchurch City Council, 2011).

Despite the implementation of the types of initiatives mentioned above, post disaster economic activity is reportedly slow and below pre-disaster levels (GoSL and UN, 2005, Colten et al., 2008). The lack of success in economic recovery initiatives can be attributed to insufficient backing from policies and legislation for employment creation and lack of consideration given to the needs of affected communities (Lewis, 2002). Kennedy et al. (2008), Pathiraja and Tombesi (2009) and Potangaroa (2009) described cases where relocation of coastal communities for risk reduction following the Indian Ocean and Samoan tsunamis resulted in the loss of traditional livelihoods such as fishing and tourism. Grant and loan schemes introduced for businesses usually come with limitations which are not attractive enough for business-owners (Mulligan and Shaw, 2007, RADA [Reconstruction and Development Agency], 2006). In Sri Lanka, there were inequities in the provision of financial and material aid to different community groups as a result of influence and politics (Lyons, 2009, Khasalamwa, 2009, Birkmann and Fernando, 2008). De Silva (2009) stated that politically powerful groups harnessed advantages of the aid programmes, and that disadvantaged people have been further fragmented and marginalised. In Indonesia, sectoral livelihood recovery programmes led to parts of the community being overlooked, showing how poor coordination and coverage of livelihood programmes inhibit recovery (Florian, 2007).

Former United States President Bill Clinton's *Key Propositions for Building Back Better* stress that a sustainable recovery process depends on reviving and expanding private economic activity and employment and securing diverse livelihood opportunities for affected populations (Clinton, 2006). Thus, the uniqueness of BBB comes from the integrated approach it proposes by giving economic recovery as much importance as reconstruction and aiming to provide solutions to suit local dynamics and preferences (Khasalamwa, 2009, Roberts, 2000, Clinton, 2006, FEMA, 2000).

Lessons learnt from post disaster experiences worldwide have provided many recommendations to improve post disaster economic recovery. Monday (2002) and the Red Cross (2010) report that the first step towards economic recovery is to obtain accurate information about the local population through data collection and consultation with local governmental authorities. A comprehensive economic recovery strategy must be created which is tailor-made to each different community based on the data obtained.

Providing assistance to businesses and entrepreneurs needs to be an essential component of the recovery process, with initiatives provided such as attractive and flexible low-interest loan packages (Bredenoord and van Lindert, 2010) and business grants and resources to support livelihoods (VBRRA, 2009). Involving local people in low-skill cash-for-work reconstruction activities is also a method to generate income while businesses and livelihoods are being re-established (GoSL, 2005, James Lee Witt Associates, 2005, Baradan, 2006).

Part of the economic recovery strategy involves considering introducing diverse and sustainable sources of income if locals are unable to continue with their previous livelihoods due to changes caused by the disaster (Olsen et al., 2005, Twigg, 2007). It is important that new livelihood options introduced must be based on locally available skills and resources and sustainable long-term (Olsen et al., 2005). Organising training programmes to support people in improving their existing livelihoods or acquiring new skills should also be part of the recovery strategy (Robinson and Jarvie, 2008, Haigh et al., 2009). Mechanisms must be put in place to monitor and support ongoing livelihood activities to maintain and strengthen economic recovery (RADA, 2006, Kennedy et al., 2008, Lyons, 2009).

Illustrations of Building Back Better: successes and opportunities

Case study: economic recovery in Sri Lanka following the 2004 Indian Ocean Tsunami

The industries most heavily impacted by the tsunami and the recovery process in Sri Lanka were fisheries and tourism, with estimated losses of US$330 million. The statistics provided in *Post-tsunami Recovery and Reconstruction: Joint Report of the Government of Sri Lanka and Development Partners* (GoSL, 2005) showed that 70–85% of tsunami-affected households regained their main source of income by November 2005 (11 months after the tsunami).

The primary disaster recovery strategy in Sri Lanka consisted of relocating communities away from the coast to minimise impacts from future coastal hazards.

The livelihood recovery programmes in Sri Lanka faced criticism for not paying attention to community needs and traditional livelihoods such as fishing, and instead encouraging the community to find new types of employment as a result of being moved away from their original coastal lands. Tourism was a major industry in coastal communities, and as a result of the new "coastal buffer zone" rule prohibiting construction along the coast, hotels were not granted permission to rebuild in the same locations. Fishermen who previously lived on the coast were being relocated 4–5 kilometres away from their original locations near the sea and therefore unable to work.

The Sri Lankan economic recovery strategy did include funding-based livelihood recovery programmes to assist the community. Cash-for-work schemes were implemented to involve locals in rebuilding. Medium- and long-term microcredit interventions by the government as well as local and international NGOs provided concessionary loans to micro, small and medium enterprises. However, due to the lack of awareness of local politics and corruption, there was inequity in the provision of funds and assistance in terms of ethnicity and gender.

NGOs with previous experience of working in Sri Lanka such as Care International and Practical Action, which were familiar with community development initiatives, understood the importance of a special focus on supporting local businesses and livelihoods at the grass-roots level during tsunami recovery operations. These NGOs initiated successful livelihood projects by providing funds and resources. In some cases, small shopping areas were built to allow merchants to start small shopping stalls, market places and boutiques. Innovative projects such as rain water harvesting in dry areas, boat building and fishery, and lagoon rehabilitation were implemented to introduce new livelihoods to the communities and support existing ones. Overall, the economic recovery experience in Sri Lanka included both BBB successes and shortcomings.

Case study: initiatives supporting economic recovery following the 2009 Victorian Bushfires

The Victorian Bushfires recovery effort provides a good BBB example, where economic recovery was considered to be a core element in the recovery strategy. The Victorian Bushfire Reconstruction and Recovery Authority (VBRRA) created an Economic Recovery Strategy along with an Economic Recovery Team within VBRRA to plan, implement and monitor economic recovery activities. The main industries in the bushfire-affected towns were agriculture, forestry and tourism. Marysville was the biggest town affected, and attracted the most tourists. As a result, Marysville consisted of many bed and breakfasts, hotels, motels, shops and several restaurant options.

The Economic Recovery Strategy included initiatives such as the provision of grants through the Victorian and Commonwealth Government business support package, low-interest flexible loans, business support and mentoring services, and tourism campaigns. The Economic Recovery Team developed and introduced new economic strategies based on observations in previous years to address issues and improve on them as recovery continued.

Grants and funds were provided for immediate restoration of businesses including A$25,000 to each directly affected business through the National Disaster Relief and Recovery Arrangements fund. Loans of up to A$200,000 were also made available at concessional interest rates. The VBRRA consulted with the local council, local Chamber of Commerce and the community recovery committees set up by the VBRRA to further identify what support businesses needed. A review in 2010–2011 highlighted that more support was needed for business recovery, and identified loans as a key driver. As a result, a new loan package was announced in July 2011 which provided a concessionary loan with 50% subsidy on the interest rate for five years.

Another good initiative by the VBRRA was to establish business support services such as the Business Information and Support Service and the Small Business Mentoring Service to provide free professional advice and counselling for business-owners. A temporary retail centre was established in Marysville to provide temporary retail spaces at highly subsidised rents while permanent locations were being rebuilt, and a "Back to Work" programme was launched to help people to return to work with assistance provided on writing resumes, job-seeking, interview preparation and accessing skills training. The physical rebuilding of places where businesses can operate from is an important factor for economic recovery. The Rebuilding Advisory Service established by the VBRRA to assist owner-builders with residential building was also available to help businesses rebuild.

A significant proportion of Marysville's income was from rental and accommodation businesses. However, two to three years into the rebuild, less than 10% of the bed and breakfasts had been rebuilt, while a majority of rental property-owners had also abandoned their properties due to the inability to rebuild. Business-owners in the affected towns had a difficult time making decisions regarding the future of their businesses due to the considerable evacuation of residents following the fires. Business-owners were hesitant to re-establish themselves in a place with a significantly reduced population. In turn, home-owners were hesitant to settle in a community with no shops and work opportunities, which resulted in a "recovery deadlock" situation. The sudden rise in construction activities also affected the local economy. The demand for building work and resources after the bushfires resulted in resource shortages and drove prices up, which affected the progress of the rebuild.

A large conference centre project was introduced in Marysville for tourism and to attract other businesses such as retail shops, gift shops and accommodation as part of the economic recovery strategy to boost the local economy. The conference centre was a milestone, and resulted in influencing people's decisions around staying or leaving.

Business rejuvenation and economic recovery posed challenges and uncertainties as a lot of the local population were displaced following the fires, but towns like Marysville did show progress over the years, with businesses re-establishing themselves. Local businesses supported each other, and the local Business Chamber of Commerce and Tourism Chambers worked together to drive economic leadership. Recovery progress was recognised, as the supermarket opened after ten months, followed by the petrol station, the local café, school and takeaway shop.

78 Economic Recovery

This was followed by the re-opening of the iconic "lolly shop", patisserie, Lake Mountain alpine resort, and Bruno's Art and Sculpture Gallery over the next few years. The new Marysville Hotel and Conference Centre opened in February 2015 and was a significant achievement and driver of the town's economic recovery. The Hotel and Conference Centre has employed 90% of its team members from the Marysville triangle region, including Marysville, Kinglake, Alexandra and surrounding towns in Murrindindi Shire which were all severely impacted by the bushfires.

Case study: planning infrastructure reconstruction to assist economic recovery in Christchurch

Christchurch, New Zealand's third largest urban city, experienced devastation from a series of earthquakes during 2010–2012. The economic toll from the earthquakes was heavy. The cost estimates of the rebuild rose from an estimated NZ$15 billion to NZ$30 billion, including the costs of business disruption, insurance administration and "Building Back Better" initiatives. The financial impact of the Canterbury Earthquakes was approximately 8% of New Zealand's GDP. The key horizontal infrastructure networks in Christchurch, including water, wastewater, stormwater and roads, were seriously affected due to the extensive land damage due to liquefaction that took place as a result of the earthquakes, which impacted businesses and industries.

One of the significant aftermaths of the Canterbury Earthquakes was the considerable damage caused to Christchurch's Central Business District (CBD). Prior to the earthquakes, the CBD constituted of approximately 6,000 businesses accounting for 25% of Christchurch's employment. Due to the large amount of unreinforced masonry and old heritage buildings designed to outdated structural standards, the earthquakes caused serious damages in the CBD. Following the major earthquakes and aftershocks in 2010 and 2011, it was found that nearly 50% of the buildings in the CBD would have to be demolished. Due to the extensive danger posed by damages in the CBD and the demolition process, a cordon was placed around the CBD preventing public access into the area, which was only completely removed in June 2013. The cordon disrupted CBD businesses, resulting in their having to relocate, and in some instances close down.

The horizontal infrastructure rebuild in Christchurch was undertaken by a newly formed alliance called the SCIRT (Stronger Christchurch Infrastructure Rebuild Team), which consisted of Christchurch City Council, Canterbury Earthquake Recovery Authority (CERA), the New Zealand Transport Agency and consulting and contracting companies including City Care, Downer, Fletcher, Fulton Hogan and McConnell Dowell. The SCIRT planned, designed and executed the rebuilding and replacement of the damaged horizontal infrastructure, and made a point to facilitate business and economic recovery during the rebuild.

The SCIRT carefully planned its rebuilding activities in alignment with other rebuilding projects in the best interests of local people. It coordinated with the rebuilding and repair projects undertaken for the electricity, gas and telecommunications networks to minimise disruption to the community who lived and worked in the areas. Projects to restore water, wastewater and stormwater services and road networks to the CBD were prioritised to encourage businesses to return as soon as possible after the CBD cordon was lifted. The SCIRT's central city programme staff consistently liaised with the Central City Development Unit (CCDU) which co-ordinated the rebuilding activities within the central city to ensure that the rebuild supported economic activities.

The SCIRT ensured compliance with CERA's recovery strategy and maintained constant communication with the business community in order to keep them informed and cater to their needs through fact sheets distributed via e-mail newsletters, letters, social media and local newspapers. The SCIRT facilitated the rebuilding process for businesses by sending maps of the infrastructure construction sequences outlining road closures and interruptions to services such as water well in advance of works, followed by reminders to ensure that businesses could prepare themselves. In establishing construction sites, the SCIRT and CCDU coordinated with businesses operating nearby to minimise disruption and enable access to businesses during construction.

Equipment used in the rebuild of Christchurch's water infrastructure was sourced from local suppliers as often as possible, to support the local economy. The Christchurch–West Melton aquifer system, which is the source of Christchurch's water supply, provides an estimated economic value of NZ$1,298 million per year to New Zealand. Renewing and upgrading water supply infrastructure during the rebuild was therefore directly funded by the central and local governments as an important component of the rebuild.

The infrastructure engineering sector, both in terms of design and construction, was struggling in the Christchurch and Canterbury area. However, the infrastructure rebuild following the earthquakes provided an opportunity for these businesses to recover and have a consistent workload for the foreseeable future. Businesses in the construction sector positioned themselves for the volume of work presented by the rebuild by increasing staff numbers to keep up with project demands by recruiting labour from other cities within New Zealand as well as overseas to manage the workload. As a result, many people from around New Zealand moved to Christchurch for better employment opportunities, thus the rebuild in Christchurch posed a good opportunity to rejuvenate and support economic growth in the construction industry in Christchurch and in New Zealand as a whole as part of Building Back Better.

Economic Recovery for Building Back Better

Successful economic recovery requires strong commitment from the government. Post disaster recovery, including economic recovery, should be tailor-made to suit each different community. In order to understand the local community, data must be collected on aspects such as people's livelihoods, skills, income levels and preferences. Post disaster practices can be made more efficient by collating this data in the pre-disaster stage. Local councils are in the best position to carry this out, and therefore should lead or be engaged in this process. Having a tailor-made economic recovery strategy leads to a successful economic recovery process. The economic recovery strategy should be based on local information, and needs to identify and support beneficiaries and entrepreneurs in particular. The strategy should empower locals to re-establish traditional livelihoods or adopt new livelihoods and should encourage the use of local resources. Coordinating the rebuilding process to assist and complement business recovery should be considered. The strategy needs to also include a metric to measure economic recovery and track progress.

The recovery of businesses is fundamental to economic recovery in a community. Funding is a key element to support business recovery. Therefore, post disaster recovery needs to consider providing funding mechanisms for businesses such as grants and flexible low-interest loans. Supporting businesses through business mentoring or business advice and counselling services can assist them in obtaining the necessary information and resources to make informed decisions during recovery. The economic recovery process should also be supported through up-skilling people based on skills shortages by arranging low-cost training programmes as part of the recovery strategy.

Economic recovery in a community is co-dependent on the recovery of the community's local residents. Resident versus business deadlocks have been identified as a real issue in post disaster environments where residents are hesitant to remain in a town with no businesses, while businesses are hesitant to re-open in towns with no residents. These deadlocks must be identified early on and dealt with by supporting businesses to re-establish themselves as well as attracting residents to stay. This process can be supported by keeping the community informed about business recovery plans, establishing temporary retail and work spaces for businesses to re-establish themselves as soon as possible, introducing new large business ventures to boost the economy, create new jobs and attract residents and tourists, and facilitate rebuilding of business premises by fast-tracking permit procedures and attracting builders. Post disaster economic recovery needs to also include advertising and marketing to promote local industries and attract new residents and tourists.

BBB Indicators for Economic Recovery

Indicators or best practices for Building Back Better for Economic Recovery were developed based on case study research findings and international examples. Economic Recovery for BBB can be grouped under Economic Recovery Strategy, Funding, Decision-making and Training, and Business Support and Promotion. The BBB Indicators for Economic Recovery are listed in Table 7.1. The Indicators serve as a practical guide to direct stakeholders involved in post disaster activities to understand the elements that need to be considered when planning and implementing reconstruction and recovery programmes in order to Build Back Better.

Table 7.1 Build Back Better Indicators for Economic Recovery

Economic Recovery Strategy	Develop a tailor-made Economic Recovery Strategy catering to local needs using information collected from locals and local councils
	Empower and support locals to re-establish traditional livelihoods and to upgrade facilities and technologies for business rejuvenation if appropriate
	Introduce new livelihood options utilising local resources and opportunities for up-skilling to cater to skills shortages
	Plan economic restoration activities concurrently with rebuilding (e.g. rebuilding of infrastructure contributing to key economic activities prioritised)
	Adopt a measurement tool to track economic recovery progress
Funding, Decision-making and Training	Provide government support to assist business recovery (e.g. provision of special government grants and flexible low-interest loans)
	Put in place business advisory services to support and advise businesses
Business Support and Promotion	Support speedy re-establishment of businesses through setting up temporary retail/work spaces for businesses, fast-tracked permit procedures and incentives provided to skilled builders to facilitate rebuilding, and fast-tracked insurance settlements
	Consideration of alternative, innovative options if economic recovery progress is poor (e.g. introducing a big business such as a conference centre, shopping mall, sports stadium) to boost the economy, create new jobs and attract residents and tourists if economic and community recovery progress is poor
	Keep the local community regularly informed on economic recovery plans and progress to encourage residents and boost morale
	Advertise and promote local industries and attractions to appeal to tourists

References

Baradan, B. (2006). Analysis of the Post-disaster Reconstruction Process following the Turkish Earthquakes, 1999. In: Group, I. R. (ed.) *International Conference on Post-Disaster Reconstruction Meeting Stakeholder Interests.* Montreal, Canada: University of Montreal.

Birkmann, J. & Fernando, N. (2008). Measuring Revealed and Emergent Vulnerabilities of Coastal Communities to Tsunami in Sri Lanka. *Disasters,* 32, 82–105.

Bredenoord, J. & van Lindert, P. (2010). Pro-poor Housing Policies: Rethinking the Potential of Assisted Self-help Housing. *Habitat International,* 34, 278–287.

Christchurch City Council (2011). *Central City Plan: Draft Central City Recovery Plan for Ministerial Approval.* Christchurch, New Zealand: Christchurch City Council.

Clinton, W. J. (2006). *Lessons Learned from Tsunami Recovery: Key Propositions for Building Back Better.* New York: Office of the UN Secretary-General's Special Envoy for Tsunami Recovery.

Colten, C. E., Kates, R. W. & Laska, S. B. (2008). Three Years after Katrina: Lessons for Community Resilience. *Environment,* 50, 36–47.

de Silva, M. W. A. (2009). Ethnicity, Politics and Inequality: Post-tsunami Humanitarian Aid Delivery in Ampara District, Sri Lanka. *Disasters*, 33, 253–273.

FEMA (2000). *Rebuilding for a More Sustainable Future: An Operational Framework*. Washington, DC: Federal Emergency Management Agency.

Florian, S. (2007). Housing Reconstruction and Rehabilitation in Aceh and Nias, Indonesia – Rebuilding Lives. *Habitat International*, 31, 150–166.

GoSL (2005). *Post-tsunami Recovery and Reconstruction: Joint Report of the Government of Sri Lanka and Development Partners*. Colombo, Sri Lanka: Government of Sri Lanka.

GoSL & UN (2005). *National Post-Tsunami Lessons Learned and Best Practices Workshop*. Colombo, Sri Lanka: Government of Sri Lanka and United Nations.

Haigh, R., Amaratunga, D., Baldry, D., Pathirage, C. & Thurairajah, N. (2009). *ISLAND – Inspiring Sri Lankan Renewal and Development. RICS Research*. Salford, UK: University of Salford.

James Lee Witt Associates (2005). *Building Back Better and Safer: Private Sector Summit on Post-tsunami Reconstruction*. Washington, DC: James Lee Witt Associates.

Kennedy, J., Ashmore, J., Babister, E. & Kelman, I. (2008). The Meaning of "Build Back Better": Evidence from Post-tsunami Aceh and Sri Lanka. *Journal of Contingencies & Crisis Management*, 16, 24–36.

Khasalamwa, S. (2009). Is "Build Back Better' a Response to Vulnerability? Analysis of the Post-tsunami Humanitarian Interventions in Sri Lanka. *Norwegian Journal of Geography*, 63, 73–88.

Lewis, C. (2002). *Desertification, Natural Disasters, Plight of Small Island Developing States Dominate Second Committee Discussion of Environment, Sustainable Development – Part 3 of 3*. Available: https://www.un.org/press/en/2002/GAEF3006.doc.htm [Accessed 9 August 2018].

Lyons, M. (2009). Building Back Better: The Large-scale Impact of Small-scale Approaches to Reconstruction. *World Development*, 37, 385–398.

Mannakkara, S. & Wilkinson, S. (2012). Build Back Better Principles for Economic Recovery: The Victorian Bushfires Case Study. *Journal of Business Continuity and Emergency Planning*, 6, 164–173.

Monday, J. L. (2002). Building Back Better: Creating a Sustainable Community after Disaster. *Natural Hazards Informer*, 3. Available: https://hazards.colorado.edu/archive/publications/informer/infrmr3/informer3b.htm [Accessed 9 August 2018].

Mulligan, M. & Shaw, J. (2007). What the World Can Learn from Sri Lanka's Post-tsunami Experiences. *International Journal of Asia-Pacific Studies*, 3, 65–91.

Okuda, K., Ohashi, M. & Hori, M. (2011). On the Studies of the Disaster Recovery and the Business Continuity Planning for Private Sector Caused by Great East Japan Earthquake. In: Cruz-Cunha, M. M., Varajão, J., Powell, P. & Martinho, R. (eds) *ENTERprise Information Systems*. Berlin, Germany: Springer.

Olsen, S. B., Matuszeski, W., Padma, T. V. & Wickremeratne, H. J. M. (2005). Rebuilding after the Tsunami: Getting It Right. *AMBIO: A Journal of the Human Environment*, 34, 611–614.

Pathiraja, M. & Tombesi, P. (2009). Towards a More "Robust" Technology? Capacity Building in Post-tsunami Sri Lanka. *Disaster Prevention and Management*, 18, 55–65.

Potangaroa, R. (2009). Native Engineering Technologies: The 2009 Samoan Tsunami and Its Significance for New Zealand. Unpublished report.

RADA (2006). *Sri Lanka: Mid-year Review – Post Tsunami Recovery and Reconstruction – June 2006*. Colombo, Sri Lanka: Reconstruction and Development Agency.

Red Cross (2010). *World Disasters Report 2010 – Focus on Urban Risk*. Geneva, Switzerland: International Federation of Red Cross and Red Crescent Societies.

Roberts, P. (2000). The Evolution, Definition and Purpose of Urban Regeneration. In: Roberts, P. & Sykes, H. (eds) *Urban Regeneration: A Handbook*. London: SAGE Publications.

Robinson, L. & Jarvie, J. K. (2008). Post-disaster Community Tourism Recovery: The Tsunami and Arugam Bay, Sri Lanka. *Disasters*, 32, 631–645.

Twigg, J. (2007). *Characteristics of a Disaster-resilient Community – a Guidance Note*. London: DFID Disaster Risk Reduction Interagency Coordination Group.

VBRRA (2009). *Nine Month Report*. Melbourne, Australia: Victorian Bushfire Reconstruction and Recovery Authority.

VBRRA (2010). *12 Month Report*. Melbourne, Australia: Victorian Bushfire Reconstruction and Recovery Authority.

8 Institutional Mechanism

Introduction

Recovery and reconstruction following a disaster is a complex and high-pressure environment. Recovery and reconstruction requires the participation of many different types of stakeholders who often have no previous experience of working together or working in post disaster environments (Lloyd-Jones, 2006). The quality and outcome of recovery depends on who is in charge, how different stakeholders work together and how the recovery effort is planned and implemented (Clinton, 2006).

This chapter describes how an appropriate institutional mechanism should be adopted for effective implementation of recovery for BBB. Governance mechanisms, multi-stakeholder management, partnerships, grass-roots-level involvement and quality assurance are covered. BBB propositions for Institutional Mechanism, which serve as practical recommendations for implementation developed through international case study research and literature review, are presented, along with case study examples demonstrating these propositions in action.

Post disaster institutional mechanisms for planning and implementing recovery

Post disaster environments are unique and often require large-scale rebuilding and development work in a short period of time involving a large number of stakeholders, both locally and internationally. Post disaster institutional mechanisms need to ensure speed and quality in the rebuilding effort while facilitating cooperation and collaboration between stakeholders to produce a unified outcome (Khasalamwa, 2009, Lyons, 2009). The institutional mechanism adopted also needs to allow for appropriate consultation with stakeholders and the community to identify needs, fast and intelligent decision-making, effective mobilisation and management of recovery funds, and transparency and accountability (Ozcevik et al., 2009).

Many reconstruction efforts have failed to successfully incorporate BBB concepts for risk reduction and social and economic recovery that can benefit the local community because of the lack of grass-roots-level involvement, poor coordination, lack of expertise, conflicting interests, poor budget controls and corruption (James Lee Witt Associates, 2005, UNISDR, 2005). Haas et al. (1977) state that the success of recovery programmes is attributed to the quality of leadership, planning and organisation. Rubin et al. (1985) reiterate that leadership is a key contributor to effective recovery, along with the ability to act, knowledge of available resources, and the capacity of local officials.

Samaratunge et al. (2012) state that the interrelationships between various actors in post disaster environments influence recovery outcomes. When centralisation or the level of internal control are too dominant, communities become less adaptive to change, thus stifling progress and innovation. On the other hand, if there is too much decentralisation, it can result in a chaotic, ad hoc environment.

One of the most common issues with post disaster environments is the difficulty in coordinating between stakeholders to produce a unified outcome (GoSL and UN, 2005). Initially, there is often no organisation in charge of the overall recovery effort (Frerks and Klem, 2005). The lack of guidance leads different stakeholders to participate disjointedly, promoting personal agendas which conflict with the interests of the local community (Batteate, 2006). The pressure for fast results during recovery also prevents well-intentioned stakeholders from considering community needs (Lyons, 2009). Ambiguity about the roles of different stakeholders is another issue (GoSL and UN, 2005). Many stakeholders involved in recovery have no previous experience in post disaster environments, leading to ad hoc responses.

Often post disaster interventions are governed by the national government without sufficient consultation or power given to local councils (Clinton, 2006, Frerks and Klem, 2005). Local-level organisations with useful local knowledge lack the capacity to operate to their full extent when impacted by disasters, and are therefore excluded from recovery efforts (Lloyd-Jones, 2006). The involvement of many different external stakeholders who have competing interests as well as lacking proper local knowledge leads to unsatisfactory outcomes for the local community (Lyons, 2009, Ruwanpura, 2009). The lack of specific role allocation of stakeholders resulting in misunderstood responsibilities and duplication of activities by different stakeholders, poor coordination between stakeholders resulting in confusion and inefficiency, and low-level involvement of local stakeholders are common issues found in post disaster reconstruction environments.

Therefore, choosing an appropriate institutional mechanism and an appropriate level of centralisation/decentralisation is very important for effective post disaster recovery. The World Bank's *Handbook for Reconstructing after Natural Disasters* (Jha et al., 2010) introduces two options for post disaster institutional mechanisms: using existing government organisations or creating a new recovery authority. The level of centralisation can vary between centralised recovery efforts and decentralised recovery efforts. The choice is whether to create a new/dedicated recovery authority or to manage recovery through existing government organisations.

Using existing government organisations

Using existing government organisations can be useful as they already have well-established processes, relationships with other departments, resources, and local knowledge and expertise. However, the success of existing organisations in a post disaster context is dependent on how well those arrangements were performing prior to the disaster. If the government organisations have been directly impacted by a disaster, they may suffer losses in capacity which would hinder their operations (Ruwanpura, 2009). Existing organisations also tend to have rigid, time-consuming procedures which are not suitable for high-pressure post disaster environments that require flexibility and fast results that are ideally required for BBB (Mora and Keipi, 2006, Ozcevik et al., 2009, Kennedy, 2009).

The 2008 Wenchuan Earthquake reconstruction process in China provided an example where existing government structures were utilised for recovery. Central government provided leadership and determined the policies, rules and regulations for rebuilding. Departments responsible for disaster reduction, relief and recovery in the State Council carried out these duties. Local governments set up coordination offices locally to handle disaster reduction, relief and recovery at the local level. As a result of this governance structure, the recovery process paid little attention to collaboration with NGOs and was unable to benefit from their services, so the social recovery of affected communities suffered.

Creating a new recovery authority

Creating a new or dedicated organisation is a common approach to reconstruction and recovery, and has been adopted in many major recovery efforts, including the 2004 Indian Ocean Tsunami recovery in 2005, the Hurricane

Katrina recovery in 2009, the Victorian Bushfires recovery in 2010, the Haiti Earthquake recovery in 2010/2011, the Canterbury Earthquakes recovery and the 2011 Great Eastern Japan Earthquake and Tsunami recovery.

Japan, which experiences frequent earthquakes, switched from its traditional government-led recovery structure to having a separate institution for the recovery process following the earthquake and tsunami in 2011. Authors such as Jha et al. (2010), Olshansky (2005) and Thiruppugazh (2014) state that having a separate institutional arrangement is more appropriate to handle the challenges present in post disaster environments, and supports BBB by improving efficiency and effectiveness. Thiruppugazh (2014) points out that having an entity solely responsible for reconstruction and recovery also sends a message to the community that there is a strong commitment to rebuild and recover. However, on the other hand, creating a new institution with the necessary resources, legislative and regulative powers and authority may be time-consuming and difficult.

When setting up a separate recovery institution, important considerations such as its mandate, structure and timeframe need to be addressed. The way in which a new body can coordinate with and manage the various local and non-local stakeholders involved in recovery is critical for recovery success. If the institution is temporary, it is necessary to determine a non-disruptive exit strategy that transfers the lessons learnt to existing organisations. Countries in which short-term recovery institutions were established suffered from loss of the knowledge gained during recovery and an absence of long-term disaster management (James Lee Witt Associates, 2005, Red Cross, 2010). If the recovery institution is to be permanent, then it is important to establish what its powers, functions and financial arrangements will be outside the reconstruction period.

Centralised recovery

Highly centralised recovery programmes display top-down control by the governing body (government or recovery authority). Recovery plans and programmes are developed and implemented with little or no participation and consultation at the grass-roots level, and do not sufficiently focus on social recovery for BBB (Davidson et al., 2007). However, this type of recovery programme is often efficient, cost-effective and completed on time (Huang et al., 2011, Johnson et al., 2006).

Despite being more efficient, centralised recovery efforts are often problematic as the lack of participation at the grass-roots level leads to unsuitable and unsustainable recovery solutions for local communities (Clinton, 2006, Frerks and Klem, 2005, Ruwanpura, 2009, Tas, 2010). Highly centralised approaches are also known to create high levels of dependency in communities, and thus do not improve their resilience (Romeo, 2002, Davidson et al., 2007).

Decentralised recovery

Decentralised recovery programmes include the community and local-level organisations in decision-making, planning and implementation of recovery projects (Monday, 2002, Clinton, 2006, Davidson et al., 2007). Access to local knowledge and indigenous resources lead to innovative solutions to recovery problems (de Silva, 2009), and the use of existing local networks and organisational structures facilitates knowledge-sharing and implementation in line with BBB Principles (Brinkerhoff, 2005, Uphoff, 2000). Community involvement also aids psychological recovery of disaster victims (Gordon, 2009). However, too much control given to communities can lead to delays in recovery planning, disputes among opposing groups within the community and a lack of emphasis given to technical aspects such as resilient building design (Mannakkara and Wilkinson, 2015, Pathiraja and Tombesi, 2009).

It is therefore important to consider the characteristics of the local community and local context to determine the most appropriate institutional mechanism to plan and implement the recovery process.

Illustrations of Building Back Better: successes and opportunities

Case study: post disaster recovery following the 2009 Victorian Bushfires

The 2009 Victorian Bushfires was the largest bushfire event in Australia. Due to the extensive damages and substantial recovery effort required, the Victorian Government created the Victorian Bushfire Reconstruction and Recovery Authority (VBRRA) as a separate recovery authority to manage the bushfire recovery. The VBRRA was set up as a coordinating body and was the central point of contact for all governmental agencies and stakeholders involved in recovery. The VBRRA developed the Recovery Strategy for the bushfire affected areas and consisted of various teams to oversee the different aspects of recovery such as temporary accommodation, donations management, the housing rebuild, social recovery and economic recovery.

Despite the VBRRA being in action, there were some challenges in the recovery process which led to important lessons being learnt about the role of a recovery authority. There was a lack in role allocation for stakeholders, which at times led to duplication of recovery activities. The VBRRA encouraged collaboration between stakeholders, but without formal mechanisms for partnerships, collaboration did not occur in practice. A recovery authority needs to allocate roles to stakeholders clearly, and encourage and foster partnerships between stakeholders using formal mechanisms such as regular meetings. It is also important to facilitate and encourage knowledge- and information-sharing between departments and organisations, along with people and resources as appropriate to work towards common goals. Being a new body, the recovery authority needs to consult local councils and utilise existing knowledge about disaster recovery from experts and other relevant sources.

The VBRRA's recovery model held the community at its core. The VBRRA organised community meetings to include the community in open discussions with regard to the recovery process. The VBRRA also set up Community Recovery Committees in each affected town to facilitate communication with the different communities. Despite the community being consulted, final decisions were made at the Victorian Government level and were sometimes not in line with community preferences. Locals did not like the modern architecture of some community buildings built during the rebuild, and new infrastructure built without due consideration of the small rate bases and low income levels of the local council created issues with long-term maintenance. Although community-centred recovery was recognised as important, there was a practical failure in implementing these concepts from top-down government structures.

Owner-building was promoted for residential reconstruction in Victoria to involve and empower the local community. Home-owners acted as the project managers for the construction of their homes, and employed small-scale builders and tradespeople. However, this process became problematic as there was a significant shortage of small-scale builders, who were already occupied with work in Melbourne and were not interested in participating in the bushfire rebuild. Although the VBRRA organised volume builders for major infrastructure projects, there was insufficient consideration given to organising builders for residential reconstruction, which affected recovery progress. The VBRRA then set up the Rebuilding Advisory Service to assist home-owners with the rebuild and to provide advice and assistance on working with builders and sub-contractors.

The involvement of different stakeholders in a major project such as post disaster reconstruction and recovery requires a high level of quality control. BBB recommendations suggest that quality assurance be implemented through monitoring, inspections and training of stakeholders. Many stakeholders in the bushfires rebuild did not have sufficient experience or training in specific post disaster skills. Changes made to the building code during the recovery process led to different builders interpreting the code in different ways due to a lack of training. As the recovery authority, the VBRRA needed to also be responsible for implementing a quality control and monitoring mechanism.

Deciding when the VBRRA should be shut down was a critical issue. The VBRRA's mandate was for two years, but was extended several times due to ongoing work and finally wound down in 2012. The Fire Recovery Unit was then created as a temporary organisation following the VBRRA, to streamline the VBRRA's remaining work back to existing government departments and the local council. The closure of the recovery authorities revealed the importance of retaining the knowledge and resources obtained during the bushfires recovery, and the need for a permanent organisation to overlook disaster management in Victoria. As a result, Emergency Management Victoria was established, which now leads emergency management in Victoria to strengthen the capacity of communities to withstand, respond to and recover from emergencies.

Despite the creation of the VBRRA as a coordinating body, better use of the VBRRA's position, powers and resources could have created better recovery outcomes. While adopting a recovery authority is a good option for efficient post disaster recovery, better understanding of the recovery authority's role and responsibilities in the recovery effort can assist in improving the outcome.

Case study: the 2004 Indian Ocean Tsunami recovery process in Sri Lanka

Due to the large-scale, far-reaching impact of the Indian Ocean Tsunami, the President of Sri Lanka appointed task forces to manage the post disaster activities. The task forces included rescue and relief, law and order logistics, and the Task Force to Rebuild the Nation (TAFREN) and the Reconstruction and Development Agency (RADA) for the rebuild. The Sri Lankan Government received extensive assistance from local and international NGOs, as well as the public and private sectors, for the rebuild. The strategy for housing reconstruction was for donors to bid for projects and select affected areas or towns. Donors either funded the rebuilding of homes, where home-owners were responsible for repairing or rebuilding their homes, or for communities that needed to be relocated away from the coast, donors built new homes and housing settlements inland. A coordinated effort was planned, with bi-weekly meetings arranged between high-level government officials and key representatives of the donor community.

There was a large influx of local and international NGOs to conduct recovery operations. The NGOs worked under pressure to achieve fast results in an uncoordinated manner, resulting in a disjointed recovery effort. The lack of local awareness and links with local authorities also resulted in impractical recovery solutions by NGOs. Large new housing settlements built for tsunami victims were left unoccupied by locals due to being built too far away from their places of livelihood, and in some instances located in flood zones. The tsunami rebuild experience illustrated the importance of the involvement of and leadership from locals and local authorities in recovery efforts involving non-local stakeholders to educate and train them on local needs and requirements.

In the case of owner-building, although owner-builders felt more satisfied with the homes they repaired or rebuilt with donor funding, owner-building sometimes led to structurally sub-standard homes. Therefore, a strategy such as owner-building needed to be supported through training, supervision and support from NGOs or other local organisations.

The complex administrative structure in Sri Lanka and the lack of stakeholder coordination impacted the rebuild. Priority Implementation Partnership (PIP) projects were launched as a result in 2008 to improve on the shortcomings seen in the tsunami recovery effort. The PIP projects aimed to develop and test a coordinated multi-stakeholder approach towards DRR-incorporated housing developments involving local authorities and key authorities concerned with housing and construction to develop a culture of collaboration and partnership.

Before the tsunami experience, Sri Lanka did not have any emergency management legislation or capacity to respond to emergency events. Following the creation of the recovery authorities TAFREN and RADA, the Sri Lanka Disaster Management Act No. 13 of 2005 was passed, which mandated the creation of a

Ministry for Disaster Management, National Council for Disaster Management and the Disaster Management Centre (DMC). The DMC is now a permanent institution leading all disaster management activities in the country. It is responsible for implementing and coordinating national and sub-national level programmes for reducing disaster risk with the participation of all relevant stakeholders.

The disaster management capacity at the grass-roots level was very low at the time of the tsunami, which led to poorly executed recovery operations. Local authorities and locals needed to be empowered by facilitating the availability of information, knowledge and resources through all tiers for decision-making. Since the tsunami, various training programmes have been introduced in Sri Lanka to build the disaster management capabilities in the country and to educate government officers as well as NGOs, technical officers and engineers, including programmes such as the *Coastal Community Resilience Training Workshop* and the guidelines on construction in disaster-prone areas training programme. The DMC is now responsible for initiating and conducting training programmes in collaboration with other relevant institutions.

The experience in Sri Lanka showed that despite the creation of dedicated recovery authorities to manage the recovery process, without proper planning, knowledge and awareness of the local context, role allocation, stakeholder coordination, grass-roots involvement, quality control and capacity building, recovery cannot be successful.

Case study: post disaster recovery in China following the 2008 Wenchuan Earthquake

The Wenchuan Earthquake struck Sichuan Province in China in May 2008, killing nearly 70,000 people. The devastating earthquake resulted in millions of people becoming homeless and damages exceeding US$20 billion.

The National Development and Reform Commission (NDRC), which is the planning ministry of the Chinese Government for large development programmes with national strategic importance, initiated the preparation of a state recovery plan following the ice and snow disaster that affected the southern areas of China in 2008. When the earthquake occurred, the NDRC took responsibility for planning the reconstruction and recovery effort following the principles of the state recovery plan that was in progress.

Reconstruction and recovery planning was begun very early on during the response and relief stage. This was helpful in gaining a better understanding of short-term, medium-term and long-term recovery needs. The early planning enabled more accurate resource allocation to meet needs at various stages of recovery.

In order to plan and implement the reconstruction and recovery programme in a collaborative and cohesive manner, the Government of China established a drafting group consisting of 40 ministries, provincial governments and state specialised institutes chaired by the NDRC. The central authorities planned for large-scale reconstruction, while line ministries in charge of specific sectors began to plan recovery in counties and cities. The results of the needs assessments carried out were evaluated by the NDRC, and finally the State Overall Planning

92 *Institutional Mechanism*

(SOP) for Post-Wenchuan Earthquake Restoration and Reconstruction was finalised and made effective by September 2008.

A management system was developed to implement the recovery tasks in each sector, with roles and responsibilities assigned to the local governments and relevant ministries. A coordination committee at the state level chaired by the minister of the NDRC ensured that all activities set by the SOP were being carried out. Mid-term and final overall evaluations were conducted to assess the completion of tasks and undertake performance assessments.

Due to the well-developed pre-existing structure and capacity of the Chinese Government and the NDRC, BBB was achieved through the use of existing organisations.

Case study: disaster management governance in Bangladesh

Bangladesh has a high vulnerability to disasters, being exposed to a range of hazards including floods, cyclones, tornadoes and river erosion on a frequent basis. The country adopted the *Hyogo Framework for Action* in 2005 and its successor the *Sendai Framework* in 2015, and has a well-established institutional mechanism and policy to respond to disasters in place.

All disaster management activities are governed by the National Disaster Management Council (NDMC), which is headed by the Prime Minister of Bangladesh. NDMC provides strategic guidance in disaster preparedness, response and recovery. A Ministry of Disaster Management and Relief was established under the NDMC, which acts as the key agency that formulates policy and with the Department of Disaster Management formed under the ministry, overseas the implementation and coordination of disaster response efforts in 29 sectoral ministries. The ministries are provided with an annual block fund to undertake relief and recovery operations. District and municipal-level disaster management committees were formed for local-level implementation.

Following the impacts of Cyclone Sidr and the resulting floods in 2007, further action was taken to strengthen Bangladesh's disaster management environment. In 2012, the Disaster Management Act was passed, outlining the country's legal framework for disaster management and adopting a broader definition of recovery covering private and public infrastructure, economy, livelihood and psycho-social aspects of disaster recovery. The *Bangladesh Climate Change Strategy and Action Plan* was released in 2009 along with funding mechanisms to support the country's disaster management efforts. In 2011, the *Sixth Five-year Plan of Bangladesh (2011–2015)* was published, outlining a roadmap for disaster-resilient sustainable development and Building Back Better.

Case study: post disaster recovery governance in Pakistan

The October 2005 Kashmir Earthquake was a devastating disaster event for Pakistan. The earthquake far outweighed the damages incurred from the types of hazards frequently experienced in Pakistan, such as floods and windstorms.

The Government of Pakistan adopted a four-pronged approach for post disaster recovery planning following a review of international best practices. The four-pronged approach initiated simultaneously included:

- Strategy and Standard Setting for Recovery Planning
- Setting up the Institutional Arrangements
- Setting in Motion Consultative Mechanisms
- Undertaking Preparatory Exercises, Surveys and Fieldwork

A Damage and Needs Assessment was initiated and carried out by the Government of Pakistan in partnership with the World Bank and Asian Development Bank immediately following the earthquake. The results of the assessment were released five weeks after the event, which enabled prompt arrangements of donor pledges to finance recovery as well as development of a strategic vision for recovery. Strategic objectives were developed for the reconstruction and recovery programme, including policy standards, strategic priorities, timeframes, identification of stakeholders, geographic delineation, and administrative and functional jurisdictions.

Concurrent with establishing the recovery strategy and standards for recovery planning, the Government of Pakistan started setting up appropriate institutional arrangements for post disaster recovery. A quick review and clarification of the pre-existing, multi-tiered and multi-sectoral institutional mandates was conducted, followed by a systemic process of developing institutional structures to manage and execute the recovery programme, create or modify legislation to assist with recovery, and identify and mobilise required staff and resources from all levels of government and private, technical and international development organisations.

Early disaster relief efforts were led by the Federal Relief Commission (FRC). Due to the magnitude of the disaster, the Earthquake Reconstruction and Recovery Authority (ERRA) was established a few weeks following the event. The ERRA was set up as a time-bound central authority under the prime minister's office to undertake residual relief, early recovery and long-term reconstruction activities. The ERRA's scope of work included strategic planning, resource mobilisation, coordination with all stakeholders, and monitoring and evaluation.

ERRA's apex body was the ERRA Council, headed by the prime minister, which provided strategic policy oversight and sustained financial management. An ERRA Board was established, chaired by the Chairman of the ERRA, with government and civil society representatives. The ERRA Board ensured that policy decisions, annual plans, programmes, projects and schemes were implemented. The ERRA's staff comprised civil servants from federal, state and provincial governments, the armed forces, civil society and national/international consultants. The staff, knowledge and relationships created with locals by the Federal Relief Commission which undertook early disaster relief activities were subsumed into the ERRA. This allowed the ERRA to align reconstruction programmes with the relief work conducted by the FRC and develop grass-roots-level relationships.

94 Institutional Mechanism

Provincial, district and state-level earthquake reconstruction and recovery authorities were also established, replicating the ERRA's programmatic planning and implementation models at lower levels. Line departments were included in the reconstruction programme to ensure that BBB policies would be continued once the ERRA's work was completed.

Different sector recovery policies were designed by an ERRA technical team with input from relevant international agencies (e.g. the World Health Organisation for health), which was then reviewed by sectoral Technical Advisory Groups in order to be adaptable for the local context. Input from sectoral implementers and local communities was also obtained, after which the policies were modified and presented to the ERRA Board and Council for finalisation and release.

Another function of the ERRA was to streamline and manage recovery finances. Having the ERRA act as a central body to manage and distribute recovery funds ensured that the funding was dedicated to reconstruction and recovery. Although the ERRA collected and distributed funds in a centralised fashion, implementation and decision-making were decentralised to support ownership-building. The ERRA created a tiered financial independence system where individual implementing agencies and affected communities were given autonomy over the types of initiatives to implement using the funding allocated. For example, provincial, district and state-level earthquake reconstruction and recovery authorities were given independence on approval of projects within centrally managed ERRA standards.

A Monitoring and Evaluation Wing (M&E Wing) was also established under ERRA at its inception. The functions of the M&E Wing were to focus on results and beneficiaries, lesson-learning, transparency and communication. Internal and external parallel systems of monitoring were put in place, with monitoring done at project level. Financial monitoring was done internally. M&E Wing teams were put in place to oversee technical aspects of reconstruction, social impact, construction monitoring etc. The M&E Wing was continued over the lifecycle of the projects, with monitoring carried out against key performance indicators. Project Implementation Coordination Units (PICUs) were established at the project level to ensure reconstruction progress remained on schedule and to maintain links with affected communities. The PICUs reported back to the relevant sections of the ERRA regularly.

Case study: post disaster recovery in Yemen

The recovery following the tropical storm that impacted regions of Yemen in October 2008 was undertaken by sectoral line ministries coordinated by the Office of the President and the cabinet. However, due to the ad hoc nature of recovery activities that was taking place in mid-2009, the government established the Reconstruction and Recovery Fund (RRF) to support and coordinate the recovery effort. The RRF was dedicated to achieving national, social, economic and humanitarian objectives using a strong executive management structure and decentralised implementation. The RRF ensured transparency in policies, supported effective communication between executives and local-level beneficiaries, prioritised flexibility, and fostered partnerships with local communities, NGOs and the private sector.

The project implementation process involved projects being selected by the RRF executive branch, which were discussed with local entities specialising in construction and infrastructure development. Feasibility studies were conducted by the local entities and presented to the RRF's review board to obtain project implementation approval. The RRF worked in coordination with local government in every aspect of recovery, holding regular meetings.

Case study: post disaster recovery governance in the Philippines

In 2010 the Philippines Disaster Risk Reduction and Management (DRRM) Act was passed. The Act authorised the establishment of a National DRRM Council (NDRRMC) comprised of the heads of 36 government agencies, and private sector and civil society representatives for policy-setting, coordinating and supervising DRRM activities, and conducting monitoring and evaluation. The DRRM Act mandated the creation of a National DRRM Plan, which then identified the National Economic and Development Authority (NEDA) as the lead agency responsible for carrying out recovery functions, with support from national government and regional line agencies, local government units (LGUs) and civil society organisations.

Due to the scope of the recovery effort resulting from the impacts of Typhoon Yolanda in 2013, the President of the Philippines deemed it necessary to create an agency that was exclusively dedicated to the post-Typhoon Yolanda recovery effort. Therefore, the Office of the Presidential Assistant for Rehabilitation and Recovery (OPARR) was established with a mandate of two years to develop an overall recovery strategy corresponding to short-, medium- and long-term recovery plans and programmes, and to unify the efforts of all the stakeholders involved in reconstruction and recovery.

The OPARR's role included coordinating with the NDRRMC, its member agencies and LGUs to formulate plans and programmes for reconstruction and recovery, proposing funding support for reconstruction and recovery, and monitoring and evaluation of recovery with the NEDA.

The OPARR created the Comprehensive Rehabilitation and Recovery Plan, which set out its reconstruction and recovery programme asserting BBB principles. The OPARR organised five agency clusters for reconstruction and recovery, covering infrastructure, resettlement, social services, livelihood and support. Each cluster was headed by a lead national government agency, coordinating with LGUs, civil society, private sector, international and local development partners, and other stakeholders. The Philippines already had a decentralised governance system where LGUs have significant decision- and policy-making authority, which aided local-level recovery.

The recovery effort was funded by the national government's own budget, and disbursed financing from existing government programmes and mechanisms to finance the LGUs. The OPARR also developed a tool called EMPATHY to monitor the progress of reconstruction. Monitoring of recovery funds and transparency were accomplished using a tool called Foreign Aid Transparency Hub (FAITH).

Case study: adopting an alliance for rebuilding infrastructure in Christchurch

Following the Canterbury Earthquakes in April 2011, the New Zealand Government worked through Canterbury Earthquake Recovery Authority (CERA) and Christchurch City Council to decide how to respond to the large infrastructure rebuild ahead. Lateral infrastructure posed a big challenge due to extensive liquefaction as a result of the earthquakes, causing damage to roads, and water and wastewater infrastructure. The Stronger Christchurch Infrastructure Rebuild Team (SCIRT) was established to provide an end-to-end service consisting of investigation, design and construction, and management of the lateral infrastructure rebuild.

The SCIRT was formed as an alliance between CERA, the New Zealand Transport Agency, Christchurch City Council, and major construction companies and consultancies in New Zealand, including Fulton Hogan, Downers, Fletcher Building, McConnell Dowell and City Care. The Horizontal Infrastructure Governance Group was established as a separate body for the three client organisations involved with the SCIRT to discuss the SCIRT's functions and matters regarding funding. The importance of using local resources was a common driver in establishing the SCIRT. Ten per cent of the initial employment in the SCIRT was to consist of unemployed people from Christchurch, and two-thirds of the work was delivered by local sub-contractors.

The scale of the rebuild meant that no single contractor had sufficient experience or resources to tackle the project, therefore it was decided that a collaborative response was required. Having the client, local council, and the contractors working together brought broadened knowledge and experience to the alliance. The different parties were able to learn from each other and work collaboratively to achieve common goals. The SCIRT also greatly sped up the design, consent and construction process. Having the SCIRT in place eliminated the need to tender, and design and consent processes were fast-tracked due to the in-house personnel and expertise. The collective effort also enabled innovative solutions to be explored.

Uncharacteristic of typical alliance arrangements, the client organisations in the SCIRT implemented some in-house competition between contractors within the SCIRT to drive down costs, whereby contractors received their share of work based on performance. The contractor's performance was assessed using a set of key performance indicators. Collaboration in the SCIRT was achieved through a payment mechanism where bonus payments were made and split between the clients and the contractor based on money saved by each contractor. However, all stakeholders belonging to the SCIRT worked with a common goal of rebuilding Christchurch as efficiently as possible.

The creation and operation of the SCIRT to manage the horizontal infrastructure recovery in Christchurch is a success story, presenting the formation of an alliance as a good option for bringing stakeholders together for effective reconstruction and recovery. The lessons learned and innovations developed throughout the SCIRT were passed on through a Learning Legacy and made available on a SCIRT Legacy website for future learning.

Institutional Mechanism for Building Back Better

Adopting a suitable institutional mechanism is crucial for Building Back Better. The chosen institutional mechanism and how it operates determines the efficiency and effectiveness of the post disaster recovery planning and implementation process.

Past disaster experiences have shown that reactive, ad hoc arrangements deliver poor results. Preparedness in setting up an institutional mechanism to plan and implement the reconstruction and recovery programme immediately after a disaster event before recovery begins, or preferably in the pre-disaster period, is needed.

A well-founded institutional delivery mechanism must be developed in order to plan and implement post disaster recovery to Build Back Better, ensuring effectiveness and efficiency. The implementing body chosen to deliver the reconstruction and recovery programme needs to be a local or national-level autonomous authority with sufficient capacity, resources, authority and legislative backing to plan, implement and monitor the reconstruction and recovery programme. The authority needs to have an over-arching view of the recovery process, including sector recovery plans and projects, project sequences, timeframes and resource allocation information. The institutional delivery mechanism has to be responsible for establishing the guiding principles for the recovery effort in line with BBB Principles, create a tailor-made post disaster reconstruction and recovery programme suitable for the local community, specify stakeholder roles and responsibilities, and put mechanisms in place to monitor and evaluate the recovery effort.

The primary decisions around choosing an institutional mechanism for post disaster recovery are (1) whether to create a new recovery authority or whether existing government organisations can be used, and (2) what level of centralisation or decentralisation is needed. The most important factor to consider in order to Build Back Better is to ensure that the institutional delivery structure chosen works in line with current local governance structures and regulatory frameworks of the affected communities in order to be effective and practical.

Existing government organisations can be used if they have sufficient capacity, resources, in-house knowledge, expertise and strong partnerships with other government organisations to plan and implement a reconstruction and recovery programme. The use of existing government organisations is appropriate where comprehensive systems for dealing with post disaster situations have already been put in place in pre-disaster periods, where they have well-established facilitation mechanisms to assist recovery, and where they have pre-existing relationships with national and local-level stakeholders.

The creation of a new or dedicated recovery authority to respond to reconstruction and recovery is the popular choice in most countries where the capacities and capabilities of existing governmental organisations are insufficient. Recovery authorities are usually created by a legislative mandate which gives the recovery authority powers to alter existing implementation mechanisms, legislation and regulation to benefit the recovery process. The newly created recovery authority is usually staffed with employees from government, non-government and private sector organisations with relevant knowledge, expertise and relationships. The recovery authority can have a short-term mandate in response to a specific disaster event, or it can be sustained as a permanent institution responsible for pre- and post disaster management functions.

In the case of a recovery authority established with a short-term mandate, an exit strategy needs to be generated to enable a smooth transition back into normal government streams of operation while providing sufficient continued support with recovery-related issues. Recovery-related information must be transferred from the recovery authority to the government organisations responsible for different sectors. A few personnel can be dedicated within each government organisation to respond to recovery-related issues following the termination of the recovery authority. The exit strategy must take into account the unpredictable nature of post disaster environments and remain flexible. It must allow ample time for reconstruction and recovery projects to be launched and carried out satisfactorily. The recovery authority must only be terminated once its specific services are no longer required.

Grass-roots-level information and knowledge need to be taken into account to plan and implement recovery programmes that are tailor-made to meet the needs of affected communities for risk reduction and social and economic recovery. Bottom-up approaches can provide valuable insights into local community vulnerabilities, sensitivities and special needs. This involves decentralisation to include local-level organisations and community members in participating and sharing their knowledge and skills. In order to be effective and efficient, the recovery needs to be coordinated and managed at a central level by the chosen implementing body.

Once the institutional mechanism is established, the next biggest challenge seen in post disaster environments is managing the multitude of stakeholders involved in reconstruction and recovery. Stakeholders need to be allocated clear roles and responsibilities. A good practice would be to register all major stakeholders involved in the recovery effort with the implementing body, with clear roles and responsibilities allocated to them to avoid confusion or duplication of tasks. It is the responsibility of the implementing body to oversee and coordinate the functions of the stakeholders to improve efficiency and effectiveness in order to Build Back Better. It is also the responsibility of the implementing body to facilitate and build partnerships and collaborations between stakeholders to improve the efficiency and effectiveness of recovery. Relationships with civil society and private sector organisations need to be strengthened. Multi-stakeholder partnerships and collaborations can be cultivated through the use of recovery plans and policies enforcing relationships for recovery projects. Effective communication between stakeholders needs to be facilitated by holding regular multi-stakeholder meetings and encouraging information-sharing. It is especially important to enable consultation between stakeholders and scientific institutions, to access technical expertise.

Funding usually poses a challenge for post disaster recovery. From past disaster experiences, BBB advocates for funds dedicated to reconstruction and recovery from government budget allocations and from donors to be directed towards a separate post disaster recovery fund. The implementing body chosen to coordinate recovery needs to be transparent to stakeholders and local communities about the funding available. Adequate funding needs to be provided to adopt resilience-building practices in all sectors in order to Build Back Better. Funding disbursement needs to be regulated and provided in stages linked with results/quality assurance to ensure that recovery projects are completed to the required standards.

Investing in creating and maintaining a comprehensive database to record all information about the recovery effort obtained from the needs assessments and data gathered during the recovery effort is necessary. The information database is an essential tool to keep track of recovery progress, manage stakeholders, manage finances, readily search for information, monitor and evaluate the recovery effort, and extract lessons learnt for legacy reports and future disaster management practices. Monitoring and evaluation of the recovery effort must be organised throughout the recovery period by providing sufficient financial and staffing resources. Lessons learnt from past experiences must be reviewed and incorporated when designing recovery programmes. The knowledge gained from a disaster needs to be preserved in the form of reports and literature, but also by creating multi-sector expert groups, including people with practical experience from participating in the reconstruction and recovery effort. These experts should be used to train and educate other staff, to expand future disaster management capacity.

BBB Indicators for Institutional Mechanism

Indicators or best practices for Building Back Better for developing a post disaster Institutional Mechanism were developed based on case study research findings and international examples. An Institutional Mechanism for BBB needs to consider Choosing an Appropriate Institutional Mechanism, Fostering Partnerships, Grass-roots-level Involvement, and Quality Assurance and Training. The BBB Indicators for Institutional Mechanism are listed in Table 8.1. The Indicators serve as a practical guide to direct stakeholders involved in post disaster activities to understand the elements that need to be considered when planning and implementing reconstruction and recovery programmes in order to Build Back Better.

Table 8.1 Build Back Better Indicators for Institutional Mechanism

Choosing an Appropriate Institutional Mechanism	Choose a local-level body (e.g. existing government organisation or new recovery authority) most suited to the local context to plan, implement and manage recovery activities, and facilitate coordination and partnerships between stakeholders involved in recovery
	Choose the level of centralisation or decentralisation most suited to the local context and community, and combine central-level coordination of recovery with decentralised planning and implementation
	If a new recovery authority is established, ensure that it includes representative local government members and local community leaders
	Devise an effective exit strategy for short-term recovery authorities to transition smoothly from rebuilding and recovery operations to business as usual
	Strengthen the capacity of disaster-impacted organisations to take part in recovery
	Establish clear roles and responsibilities for all stakeholders
	Create an exclusive, singular fund for disaster recovery to direct all funds dedicated to recovery into one pool for clear accountability and allocation for different recovery needs
	Maintain a thorough information database
Fostering Partnerships	Foster effective partnerships, collaborations and effective communication between stakeholders. Create formal partnerships if appropriate (e.g. alliances, public-private partnerships)
	Hold regular multi-stakeholder meetings to discuss, plan and implement recovery activities, avoiding duplication and allowing consultation and information exchanges between stakeholders
	Generate systems for easy and transparent access to information between stakeholders
Grass-roots-level Involvement	Emphasise decentralisation and grass-roots involvement in recovery planning and implementation
	Provide transparent information to the community
	Support local councils to take a lead role in recovery
	Involve the community as appropriate in planning, decision-making and implementation of recovery projects
Quality Assurance and Training	Use qualified, reputable stakeholders for recovery activities
	Provide training prior to recovery work as appropriate
	Put in place mechanisms for monitoring and evaluation of recovery progress
	Put in place mechanisms to transfer lessons learnt to local and national government and all relevant stakeholders for capacity-building, future resilience work, pre-disaster planning, emergency management and post disaster recovery

References

Batteate, C. (2006). Urban Disaster Risk Reduction and Regeneration Planning: An Overview. *Focus: Journal of the City and Regional Planning Department*, 3, 11–17.

Brinkerhoff, D. W. (2005). Rebuilding Governance in Failed States and Post-conflict Societies: Core Concepts and Cross-cutting Themes. *Public Administration and Development*, 25, 3–14.

Clinton, W. J. (2006). *Lessons Learned from Tsunami Recovery: Key Propositions for Building Back Better*. New York: Office of the UN Secretary-General's Special Envoy for Tsunami Recovery.

Davidson, C. H., Johnson, C., Lizarralde, G., Dikmen, N. & Sliwinski, A. (2007). Truths and Myths about Community Participation in Post-disaster Housing Projects. *Habitat International*, 31, 100–115.

de Silva, M. W. A. (2009). Ethnicity, Politics and Inequality: Post-tsunami Humanitarian Aid Delivery in Ampara District, Sri Lanka. *Disasters*, 33, 253–273.

Frerks, G. & Klem, B. (2005). *Tsunami Response in Sri Lanka: Report on a Field Visit from 6–20 February 2005*. Wageningen, The Netherlands: Wageningen University and Clingdael University.

Gordon, R. (2009). Community Impact of Disaster and Community Recovery. In: *InPsych*. Melbourne, Australia: Australian Psychological Society.

GoSL & UN (2005). *National Post-tsunami Lessons Learned and Best Practices Workshop*. Colombo, Sri Lanka: Government of Sri Lanka and United Nations.

Haas, J. E., Kates, R. W. & Bowden, M. J. (1977). *Reconstruction Following Disaster*. Cambridge, MA, MIT Press.

Huang, Y., Zhou, L. & Wei, K. (2011). 5.12 Wenchuan Earthquake Recovery: Government Policies and Non-governmental Organizations' Participation. *Asia Pacific Journal of Social Work and Development*, 21, 77–91.

James Lee Witt Associates (2005). *Building Back Better and Safer: Private Sector Summit on Post-tsunami Reconstruction*. Washington, DC, James Lee Witt Associates.

Jha, A. K., Barenstein, J. D., Phelps, P. M., Pittet, D. & Sena, S. (2010). *Safer Homes, Stronger Communities: A Handbook for Reconstructing after Natural Disasters*. Washington, DC: World Bank. Available: https://openknowledge.worldbank.org/handle/10986/2409 [Accessed 9 August 2018].

Johnson, C., Lizarralde, G. & Davidson, C. H. (2006). A Systems View of Temporary Housing Projects in Post-disaster Reconstruction. *Construction Management & Economics*, 24, 367–378.

Kennedy, J. (2009). Disaster Mitigation Lessons from "Build Back Better" following the 26 December 2004 Tsunamis. In: Ashmore, J., Babister, E., Kelman, I. & Zarins, J. (eds) *Water and Urban Development Paradigms*. London: Taylor & Francis.

Khasalamwa, S. (2009). Is "Build Back Better" a Response to Vulnerability? Analysis of the Post-tsunami Humanitarian Interventions in Sri Lanka. *Norwegian Journal of Geography*, 63, 73–88.

Lloyd-Jones, T. (2006). *Mind the Gap! Post-disaster Reconstruction and the Transition from Humanitarian Relief*. London: Royal Institution of Chartered Surveyors.

Lyons, M. (2009). Building Back Better: The Large-scale Impact of Small-scale Approaches to Reconstruction. *World Development*, 37, 385–398.

Mannakkara, S. & Wilkinson, S. (2015). Supporting Post-disaster Social Recovery to Build Back Better. *International Journal of Disaster Resilience in the Built Environment*, 6, 126–139.

Monday, J. L. (2002). Building Back Better: Creating a Sustainable Community after Disaster. *Natural Hazards Informer*, 3. Available: www.colorado.edu/hazards/publications/informer/infrmr3/informer3b.htm [Accessed 9 August 2018].

Mora, S. & Keipi, K. (2006). Disaster Risk Management in Development Projects: Models and Checklists. *Bulletin of Engineering Geology and the Environment*, 65, 155–165.

Olshansky, R. B. (2005). How Do Communities Recovery from Disaster? A Review of Current Knowledge and an Agenda for Future Research. *46th Annual Conference of the Association of Collegiate Schools of Planning*. Kansas City, MO.

Ozcevik, O., Turk, S., Tas, E., Yaman, H. & Beygo, C. (2009). Flagship Regeneration Project as a Tool for Post-disaster Recovery Planning: The Zeytinburnu Case. *Disasters*, 33, 180–202.

Pathiraja, M. & Tombesi, P. (2009). Towards a More "Robust" Technology? Capacity Building in Post-tsunami Sri Lanka. *Disaster Prevention and Management*, 18, 55–65.

Red Cross (2010). *World Disasters Report 2010 – Focus on Urban Risk*. Geneva, Switzerland: International Federation of Red Cross and Red Crescent Societies.

Romeo, L. (2002). Local Governance Approach to Social Reintegration and Economic Recovery in Post-conflict Countries: Towards a Definition and Rationale. In: *Workshop on a Local Governance Approach to Post-Conflict Recovery*. New York: Institute of Public Administration and United Nations Development Programme.

Rubin, C., Saperstein, M. D. & Barbee, D. G. (1985). *Community Recovery from a Major Natural Disaster. Programme on Environment and Behaviour Monograph #41*. Boulder, CO: Institute of Behavioural Science, University of Colorado.

Ruwanpura, K. N. (2009). Putting Houses in Place: Rebuilding Communities in Post-tsunami Sri Lanka. *Disasters*, 33, 436–456.

Samaratunge, R., Coghill, K. & Herath, H. M. A. (2012). Governance in Sri Lanka: Lessons from Post-tsunami Rebuilding. *South Asia: Journal of South Asian Studies*, 35, 381–407.

Tas, M. (2010). Study on Permanent Housing Production after 1999 Earthquake in Kocaeli (Turkey). *Disaster Prevention and Management*, 19, 6–19.

Thiruppugazh, V. (2014). Post-disaster Reconstruction and Institutional Mechanisms for Risk Reduction: A Comparative Study of Three Disasters in India. In: *Disaster Recovery*. New York: Springer.

UNISDR (2005). *Hyogo Framework for Action 2005–2015: Building the Resilience of Nations and Communities to Disasters*. Kobe, Japan: United Nations Office for Disaster Risk Reduction.

Uphoff, N. (2000). Understanding Social Capital: Learning from the Analysis of Experience of Participation. In: Serageldin, I. & Dasgupta, P. (eds) *Social Capital: A Multifaceted Perspective*. Washington, DC: World Bank.

9 Legislation and Regulation

Introduction

Most disasters are followed by legislative changes, emergency legislation and new legislative measures to assist with post disaster relief and recovery. Legislative changes can sometimes slow recovery as measures are put in place to enhance and tighten building codes and land-use, but they can also facilitate the reconstruction and recovery process to create a resilient post disaster environment.

This chapter illustrates how legislation and regulation can support BBB through ensuring compliance with BBB-based concepts and improve recovery efficiency through facilitating and fast-tracking legislative procedures. BBB Indicators for Legislation and Regulation, which serve as practical recommendations for implementation developed through international case study research and literature review, are presented.

Post disaster legislation challenges

In the post disaster environment, there is an ongoing contest between the strict application of regulations to minimise or prevent the recurrence of new and pre-existing vulnerabilities, and enabling affected communities to move back to their former habitations as quickly as possible. The more rapidly communities return to habitability, the quicker a sense of normality can be restored. These conflicting agendas often create challenges in planning and implementing post disaster recovery programmes.

After damage assessments and evaluations, building and environmental legislation should not present impediments to reconstruction and rebuilding programmes. Yet the post disaster recovery literature indicates that the implementation of certain aspects of post disaster legislation hinders the realisation of reconstruction objectives and slows the rebuild. For instance, building regulations could become burdensome in rehabilitation and reconstruction projects (Freeman et al., 2003, Martín, 2005). Excessive rules, regulations and red tape (statutory procedures) add unnecessarily to time and costs. Listokin and Hattis (2005) provide a useful analysis of two kinds of barriers that building codes could pose to rebuilding works.

"Hard barriers" are impediments to rebuilding as a result of over-regulation, without adding appreciable building value or public safety – for example, insisting on expensive structural solutions in a highly hazardous zone, where a simple alternative would be to restrict development in that zone. These types of barriers discourage rebuilding because they are added burdens.

"Soft barriers" are administrative requirements that require extra time, money and effort for reconstruction projects. These are red tape (bureaucratic procedures) that could delay new constructions and the repairing/rebuilding of damaged buildings and infrastructure (May, 2005). May (2005) suggests three sources of impediments by regulatory processes: regulatory approvals, where there are delays with consent processes and approvals, regulatory enforcement strategies and practices which are overly rigid, and unclear or confusing administrative arrangements resulting from duplication of administrative structures and gaps in regulatory decision processes. Regulatory process barriers could also result from administrative conflicts within and among disaster agencies (Listokin and Hattis, 2005). For example, rivalries between responding agencies are not uncommon among emergency services, and are an obstacle to effective emergency management.

The lack of enforcement of hazard-related laws and adequate risk-based building controls contributed to the large-scale devastation caused by the 2004 Indian Ocean Tsunami (Mulligan and Shaw, 2007). The same was witnessed in countries such as Pakistan, Turkey, Samoa and Haiti. Rebuilding without building controls further exacerbated and re-created the vulnerabilities in the built environment. Minimal consultation and consideration of the lifestyle, livelihood and economic structure of local communities when creating recovery policies can lead to increased vulnerability. The newly regulated compulsory resettlement following the tsunami resulted in the loss of traditional livelihoods of locals in Sri Lanka, and in Aceh, Indonesia the lack of regulation around the distribution of funding led to inequality and uncertainty (Khazai et al., 2006, Florian, 2007).

Lack of awareness and understanding of new legislation can also lead to non-compliance. In the post-tsunami recovery effort in Sri Lanka, external non-governmental organisations which took part did not comply with local standards due to lack of awareness (Boano, 2009). The *National Post-tsunami Lessons Learned and Best Practices Workshop* held in Sri Lanka highlighted the importance of training stakeholders (especially external NGOs) about existing and newly introduced legislation and regulations (GoSL and UN, 2005). The community's support can also be obtained by educating them about legislation and regulations that must be adhered to in reconstruction and recovery (Batteate, 2006).

Time-consuming procedures, insufficient resources to process permits and the lack of fast-tracked methods can delay reconstruction. Therefore, fast-tracked consenting procedures, collaboration with other local councils and open access to information between stakeholders are viable options to speed up recovery. Legislation can be used to remove unnecessary bureaucracy to facilitate recovery activities. It was reported that legislative suspensions and emergency powers greatly reduced highway reconstruction time following the 1994 Northridge Earthquake in the USA. Similarly, during recovery from the Australian bushfires, planning and building permits were exempted for temporary accommodation to speed up access. Planning permits for permanent dwellings were also exempted for pre-existing bushfire-affected properties that previously had planning permits, to speed up rebuilding (DPCD, 2013).

It is clear that legislation and regulatory requirements can have significant influence on the rate of recovery after a disaster event. The overall desire is for legislation to enhance the recovery and reconstruction process so that it improves the functioning of an affected community and reduces risks from future events. Any legislation changes that need to be made post disaster may be better considered before a disaster so that their implementation can be facilitated early on during recovery. Often the opportunities to introduce mitigating measures become limited over the course of recovery because of the desire to return to normalcy and thus rebuild quickly after disasters.

Well-articulated and implemented legislation should not only provide an effective means of reducing and containing vulnerabilities, but also become a means of facilitating better-thought out and designed reconstruction programmes. The main focus of recovery is the community, thus locals must be consulted and included in the process of developing new legislation and regulations to ensure the changes are suitable and beneficial for the community (Ingram et al., 2006).

Illustrations of Building Back Better: successes and opportunities

Case study: emergency management legislation in New Zealand

Prior to the 2010/2011 Canterbury Earthquakes in New Zealand, several issues were identified in relation to planning and construction legislation that could impede reconstruction of Wellington following a possible major earthquake.

Much of the legislation – in particular the Resource Management Act (RMA) and the Building Act (BA) – that existed during the period was neither drafted to cope with an emergency situation nor developed to operate under the conditions that would prevail in the aftermath of a severe seismic event. For example, the consultation procedure in the RMA could prevent meeting the reconstruction requirements within a reasonable time period.

Following the Canterbury Earthquakes, changes to both the RMA and BA were made based on recommendations by the Royal Commission set up to review the Canterbury recovery programme. Changes to seismic codes, strengthening of existing buildings, and land-use and planning parameters were made as a response to the Canterbury disaster.

Considering New Zealand before the Canterbury Earthquakes, there was an emphasis on readiness and response activities, with little consideration given to planning for sustained recovery activities. Where recovery was considered, it was over the short term, as evident in emergency awareness campaigns that encouraged communities to prepare for three to seven days after an event. Recent emergency events clearly show that longer-term recovery plans and more robust legislation are required, given the complexities associated with the rebuilding of damaged built assets.

The Civil Defence Emergency Management Act 2016 strengthens the country's planning mechanisms for recovery in the pre-disaster period itself. The Act highlights the focus on recovery by mandating local civil defence and emergency management groups and local authorities to plan for recovery in their respective areas. Recovery planning is to consider a range of scenarios depending on the types of hazards and risks in different areas, and how they will impact local communities and land.

Case study: the 1994 Northridge Earthquake recovery

The Northridge Earthquake struck Southern California in the early hours of 17 January 1994 with a magnitude of 6.8 on the Richter scale – small compared to other earthquakes, but causing significant damage.

The earthquake damaged 27 bridges and resulted in the collapse of sections of six freeways; 450 public buildings suffered significant damage, along with 6,000 commercial buildings, 49,000 housing units in 10,200 buildings suffered serious structural damage, and 388,000 housing units in 85,000 buildings had minor damage. The total value of damage to houses in Los Angeles was estimated to be about US$1.5 billion.

The Northridge Earthquake caused a shift in emphasis from disaster preparation and relief to recovery. This shift largely resulted in the success of emergency management programmes for the restoration of the affected areas. Reconstruction activities contributed to the economic revitalisation of the affected area, as the efficiency and effectiveness of response became primary objectives which were tackled by every government agency. Several bureaucratic requirements were suspended to pave the way for rapid rebuilding of damaged infrastructure.

While the rapid recovery experienced after the Northridge Earthquake can be attributed to other factors, such as political will, public policy changes and enabling, emergency management legislation played a substantial role in the rebuilding programmes after the earthquake.

Case study: Hurricane Katrina recovery

Hurricane Katrina was a Category 3 storm that struck New Orleans and the Gulf Coast on 29 August 2005. The storm surge caused severe destruction along the Gulf coast from central Florida to Texas. The most severe damage occurred in New Orleans, because of the failure of the levee system that was designed to contain storm surges, resulting in extensive flooding. The floods led to one of the largest evacuation of citizens within the US in recent times, with an estimated 1.2 million people evacuated before the incident and another 100,000–120,000 afterwards. About 350,000 houses were destroyed and over 200,000 persons required temporary shelters scattered around 16 states in the US. Hurricane Katrina was a catastrophic event with economic loss estimates of about US$200 billion.

The impacts from the hurricane were exacerbated due to insufficient foresight, planning, regulations and disaster response capacity. There were no pre-established administrative procedures, which impeded the response process, and the top-down (command-and-control) approach used posed issues. The American Bar Association in its review of the emergency system that operated during the disaster recommended more proactive response arrangements that originate from the bottom up, with local authorities as first responders.

There was also a lack of decisive leadership in the Katrina response activities. The disaster response was slow as a result of the lack of understanding of functions and roles by emergency management officials. Emergency management objectives were unclear. The US Federal Emergency Management Agency has had a confused set of objectives, largely because of its shift in focus from natural disasters to antiterrorism activities since 11 September 2001.

However, the Katrina experience led to policy changes and legislative reviews to address the problems encountered. Changes were made to the building codes in New Orleans with a view to improving the resilience of built spaces. For example, revisions were made to increase the base flood elevation levels for new construction to 3 feet or higher. This risk mitigation strategy has been tied to flood insurance cover, so that only buildings that meet the new guidelines can qualify for flood insurance and subsequent compensation. Funding sources and budget priorities have also been developed for flood protection in New Orleans.

The Katrina event necessitated the review and updating of Louisiana and New Orleans response strategies and emergency operations plans. The legislative reviews included the adoption of an all-hazards approach, expanding the scope and magnitude of anticipated hazards and allowing greater involvement of non-agency actors which proved crucial to response and recovery after the event. Partnering with non-governmental stakeholders was a paradigm shift that emerged out of the Katrina experience. Changes to land-use planning and zoning were also made to reduce the vulnerability of the New Orleans region to future flooding disasters.

The rate at which recovery is achieved is tied to the speed of reconstruction guaranteed by legislative and regulatory changes. Hurricane Katrina provided a good learning experience to emphasise the importance of legislation and regulation for BBB.

Case study: post disaster legislation for 2009 Victorian Bushfires recovery

One of the first steps taken in Australia following the bushfires was to publish a revised edition of the Australian Building Code for Bushfire-prone Areas (AS:3959) on 11 March 2009, four days following the fires. The revisions introduced six new Bushfire Attack Levels (BALs) to identify the bushfire risk of properties, and corresponding stringent design and construction requirements for greater fire protection. However, the unavailability of necessary building materials to comply with the new building code specifications and significant cost increases in the high-risk BAL zones created major delays in recovery.

Another key change in legislation concerned land-use. Soon after the fires, the entire state of Victoria was declared bushfire-prone and placed under the Wildfire Management Overlay (WMO), which meant both a planning permit and building permit were required for construction. By 2011, more accurate mapping of bushfire risk in Victoria was being carried out to replace the WMO with a Building Management Overlay (BMO). BMOs integrated the WMOs with building controls. The new planning regulations caused issues for people who had properties now included in the high-risk zones as a result of the re-mapping. This posed challenges for land-owners, as they were either unable to rebuild due to the new regulations or had to comply with costly, stringent building standards.

The processing and issuing of planning and building permits were identified as potential bottlenecks due to the high demand and low capacity in the councils of affected areas. Efforts were made to facilitate permit procedures over the recovery period. Planning permits were exempted for temporary accommodation so that emergency accommodation could be constructed quickly. Planning permits for rebuilding were exempted, and building permits were relaxed and simplified to facilitate the rebuild. However, people misunderstanding this exemption, and believing they could build without both planning and building permits was a common problem, resulting in the construction of sub-standard homes. Educating the public and stakeholders about new regulations and changes can prevent such issues.

The introduction of a "buy-back scheme" posed a solution for people on high-risk lands. Those who lost their principle place of residence in the bushfires within 100 metres of significant forest and had not started rebuilding could voluntarily apply for the scheme to buy their property. The scheme being voluntary, however, resulted in scattered uptake, leaving empty lots of land among occupied lands which needed to be maintained by the government. A compulsory buy-back scheme or a land-swap scheme could have been preferable in this instance.

The Victorian Bushfire Appeal Fund was established by the Victorian Government following the fires to provide financial support to bushfire-affected individuals and communities. The lack of regulations around the use of funds led recipients to misspend the money provided, which was an oversight. Placing a covenant on how the money should be spent could be a solution.

It was notable that there were no special permit facilitations put in place for businesses to support economic recovery. The slow revival of businesses in the affected towns impacted overall recovery. Putting in place special provisions to enable businesses to re-establish themselves speedily would benefit disaster-affected communities.

The Victorian Bushfires recovery effort adopted many good legislative and regulatory initiatives to support BBB, and shortcomings due to a lack of experience resulted in post disaster issues that can be taken as lessons for the future.

Legislation and Regulation for Building Back Better

The post disaster period is a critical time for BBB to embed resilience in affected communities by reducing vulnerabilities and improving local conditions. At the same time, communities need to be supported to return to a sense of normality as soon as possible by ensuring a smooth and efficient reconstruction and recovery process. Post disaster legislation is required to ensure compliance with BBB Principles such as Structural Resilience, Multi-hazard-based Land-use planning, Early Warning and DRR Education, Psychological and Social Recovery and Economic Recovery. Post disaster legislation also needs to be used effectively to facilitate reconstruction and recovery activities by removing bottlenecks and simplifying processes without impacting safety.

Firstly, legislative provisions are required to set up an appropriate institutional mechanism to plan and implement the reconstruction and recovery effort. This could include empowering and setting up existing governmental organisations or arrangements with special powers to enable planning and implementing the recovery effort, or providing legislative provisions to establish a new recovery authority to manage recovery.

With disaster risk reduction through improving structural resilience and multi-hazard-based land-use planning being key for Building Back Better, it is necessary to provide legislative provisions to enforce new or revised building codes and building regulations to ensure a good-quality rebuild. However, legislative and regulatory changes need close examination to avoid issues posed by resource constraints, raised costs and impacts on livelihood that can unnecessarily hinder recovery progress. Legislative provisions also need to enforce new or revised planning regulations in conjunction with building regulations to minimise hazard and risk exposure. If relocation is opted for, careful decisions need to be made on voluntary versus compulsory land buy-back schemes or land-swap schemes. In resettlement involving land-swaps, new or upgraded subdivisions need to be located in nearby low-risk areas. New subdivisions should be attractive to home-owners, with new or upgraded infrastructure, and livelihood and recreational opportunities. Community input is needed to ensure community needs are understood and met to the highest degree. Legislative provisions should not be limited to post disaster reconstruction, and need to enforce risk management and retrofitting programmes for ongoing management of hazard risks.

For BBB, legislative provisions should also be used to mandate community-inclusive and participatory recovery planning and implementation. The current literature does not attribute high importance to putting in place legislation to implement aspects such as social recovery and community consultation. Community consultation was encouraged heavily in the Victorian Bushfires rebuild. Mandating community-inclusive recovery planning and decision-making through legislative provisions will be central to creating communities that cater to local needs. The creation of community consultation groups, introducing mechanisms to allow community feedback into decision-making and holding regular multi-stakeholder meetings to include the community throughout the recovery process can be introduced through legislation. Similarly, legislation also needs to assist economic recovery by providing special provisions for businesses to rebuild and re-establish themselves.

Quality control significantly impacts the success of reconstruction and recovery and BBB. Legislation could be used for better management of stakeholders through fast-tracked tender processes for stakeholder selection, imposing quality controls, creating partnerships and role allocation/modification. The Australian case raised an issue about inconsistent contracts used by builders in residential rebuilding. It is beneficial to consider producing a standard post disaster construction contract to be used by all builders for residential building, and to provide legislative provisions to enforce it.

Recovery continues long after rebuilding has been completed, until the community reaches economic and social stability equal to or better than the pre-disaster condition. Therefore, legislation must impose continued monitoring and enforce necessary facilitations to support recovery-related activities for long-term recovery.

Legislation and regulation also need to be used to facilitate the recovery process. Simplification of permit procedures will speed up recovery, as well as encouraging adoption. Legislative provisions should be introduced to simplify and fast-track permit procedures for rebuilding. Legislative provisions can be used to expedite the release of state lands for temporary housing and resettlement operations and expedite the disbursement of funds.

Legislative provisions need to be put in place to assist business recovery, as this is often overlooked. Economic recovery should be supported through special facilitations for businesses above and beyond what are normally provided for residential rebuilding. Rebuilding of business buildings and providing more public access to businesses can be supported through special fast-tracked processes, simplified permits and subsidised resources for construction. Further assistance can be provided to set up or reinstate businesses by allowing legislative provisions for subsidised equipment, low-interest business loans and special arrangements between businesses to support each other.

There can be challenges in implementing new and revised post disaster legislation due to a lack of understanding and misinterpretation. Therefore, it is suggested that training and education should be provided for all stakeholders and the community about legislative changes.

Post disaster experiences have shown that there needs to be some flexibility about the end dates of legislative provisions. Recovery is highly dependent on the psychological state of the community and their ability to move forward. In the case of recovery operations taking longer than expected, it is important to be flexible in terms of legislative provisions.

BBB Indicators for Legislation and Regulation

Indicators or best practices for Building Back Better for post disaster Legislation and Regulation were developed based on case study research findings and international examples. Post disaster Legislation and Regulation can be used to promote Compliance with and Facilitation of BBB Principles. The BBB Indicators for Legislation and Regulation are listed in Table 9.1. The Indicators serve as a practical guide to direct stakeholders involved in post

Table 9.1 Build Back Better Indicators for Legislation and Regulation

Compliance	Put in place legislation and regulations as appropriate to enforce and comply with BBB-based recovery decisions:
	Enforce an institutional mechanism for post disaster reconstruction and recovery (existing organisations or a new recovery authority)
	Enforce updated building codes and building regulations
	Enforce updated planning regulations and land-use plans
	Enforce risk management and retrofitting programmes for ongoing hazard risk management
	Mandate community-inclusive and participatory recovery planning and implementation
	Impose quality control specifications for stakeholder selection and stakeholder management
	Enforce standardised post disaster building contracts for residential rebuilding
	Impose long-term monitoring of recovery
Facilitation	Put in place legislation and regulations to facilitate business-as-usual processes to improve efficiency for recovery:
	Simplify and fast-track permit procedures for rebuilding
	Expedite the release of state lands for temporary housing and resettlement
	Expedite disbursement of funds
	Assist business recovery
	Maintain flexibility in the end dates of legislative provisions
	Provide training and education for stakeholders and the community on new legislative changes

disaster activities to understand the elements that need to be considered when planning and implementing reconstruction and recovery programmes in order to Build Back Better.

References

Batteate, C. (2006). Urban Disaster Risk Reduction and Regeneration Planning: An Overview. *Focus: Journal of the City and Regional Planning Department*, 3, 11–17.

Boano, C. (2009). Housing Anxiety and Multiple Geographies in Post-tsunami Sri Lanka. *Disasters*, 33, 762–785.

DPCD (2013). *List of Amendments to the Victoria Planning Provisions*. Victoria, Australia: Victoria State Government Department of Planning and Community Development.

Florian, S. (2007). Housing Reconstruction and Rehabilitation in Aceh and Nias, Indonesia – Rebuilding Lives. *Habitat International*, 31, 150166.

Freeman, P., Martin, L., Linnerooth-Bayer, J., Mechler, R., Pflug, G. & Warner, K. (2003). Disaster Risk Management: National Systems for the Comprehensive Management of Disaster Risk Financial Strategies for Natural Disaster Reconstruction. Washington, DC: Inter-American Development Bank.

GoSL & UN (2005). *National Post-tsunami Lessons Learned and Best Practices Workshop*. Colombo, Sri Lanka: Government of Sri Lanka and United Nations.

Ingram, J. C., Franco, G., del Rio, C. R. & Khazai, B. (2006). Post-disaster Recovery Dilemmas: Challenges in Balancing Short-term and Long-term Needs for Vulnerability Reduction. *Environmental Science and Policy*, 9, 607–613.

Khazai, B., Franco, G., Ingram, J. C., de Rio, C. R., Dias, P., Dissanayake, R., Chandratilake, R. & Kanna, S. J. (2006). Post-December 2004 Tsunami Reconstruction in Sri Lanka and Its Potential Impacts on Future Vulnerability. *Earthquake Spectra*, 22, S829–S844.

Listokin, D. & Hattis, D. B. (2005). Building Codes and Housing. *Cityscape*, 8, 21–67.

Martín, C. (2005). Response to "Building Codes and Housing" by David Listokin and David B. Hattis. *Cityscape*, 8, 253–259.

May, P. J. (2005). Regulatory Implementation: Examining Barriers from Regulatory Processes. *Cityscape*, 8, 209–232.

Mulligan, M. & Shaw, J. (2007). What the World Can Learn from Sri Lanka's Post-tsunami Experiences. *International Journal of Asia-Pacific Studies*, 3, 65–91.

10 Monitoring and Evaluation

Introduction

The effectiveness and efficiency of post disaster reconstruction and recovery activities is crucial to the success of a community's restoration following the impact of a disaster event (Clinton, 2006). Having knowledge of Build Back Better concepts in designing recovery programmes is insufficient without systems in place to oversee and monitor implementation.

This chapter describes the importance of monitoring and evaluation throughout the recovery process to identify issues and respond as recovery proceeds, and to extract lessons to improve future disaster management practices. BBB Indicators for Monitoring and Evaluation, which serve as practical recommendations for implementation are presented.

Post disaster monitoring and evaluation

The creation of a recovery strategy to administer and guide the post disaster reconstruction and recovery effort is a common response following disaster events (GoSL, 2005, Meigh, 2009). Despite having recovery strategies and revisions in legislation and regulation to improve recovery activities, the lack of properly trained professionals competent in post disaster environments and disaster management activities can poorly affect the outcome of recovery efforts (Disaster Relief Monitoring Unit of the Human Rights Commission of Sri Lanka, 2006, Tas, 2010). Insufficient effective information- and knowledge-sharing and dissemination are also primary reasons for unsatisfactory disaster management practices (Kaklauskas et al., 2009, Lloyd-Jones, 2006). Long-term recovery beyond the reconstruction phase often does not take place due to a lack of mechanisms and expertise present, which prevents affected communities from satisfactorily Building Back Better in the long run (Business Civic Leadership Center, 2012).

The role of monitoring and evaluation in post disaster reconstruction and recovery and BBB is twofold. Firstly, monitoring and evaluation can be used to ensure compliance of recovery activities in accordance with the recovery strategy in place and relevant guidelines and regulations. The literature advocates for the need to monitor the quality and compliance of ongoing recovery efforts at a local level (Clinton, 2006, Red Cross, 2010, Colten et al., 2008, Halvorson and Hamilton, 2010). The 2003 Bam Earthquake reconstruction provides a good example where rebuilding was monitored by providing construction supervision, which assisted in assuring the quality of the rebuild (Omidvar et al., 2010). Clinton (2006) stated that the Tsunami Recovery Impact Assessment and Monitoring System (TRIAMS) was put in place during the Indian Ocean Tsunami recovery for the worst-affected countries. The recovery strategy in Christchurch, New Zealand has also been equipped with monitoring mechanisms (CERA, 2012).

Measuring recovery outputs is a good strategy for BBB (Clinton, 2006, Chang, 2010, Quarantelli, 1999). Progress needs to be monitored in a quantitative form, such as through the number of houses or buildings built and the number of businesses re-established, to ensure that the present recovery efforts are providing positive outcomes. A measure of the wellbeing/social recovery of people is also required. Clinton (2006) also suggested that long-term recovery should be monitored through continued data collection to ensure that recovery efforts do not leave communities with residual issues.

Secondly, monitoring and evaluation can be used to evaluate recovery efforts to obtain lessons for the future and improve future disaster management and post disaster reconstruction and recovery efforts. Proper monitoring can help to identify problems with post disaster interventions, to rectify them as well as obtain lessons (Monday, 2002, Wiles et al., 2005). These lessons need to be incorporated into revised policy and procedures for future disaster management practices (FEMA, 2000, Grewal, 2006, Tas, 2010). Japan displayed successful examples of incorporating lessons learnt into practice where infrastructure systems and building regulations were improved based on previous disaster experiences (Matanle, 2011, Norio et al., 2011). Disaster management concepts learned from the Indian Ocean Tsunami experience were incorporated into the education systems in Sri Lanka and India (DN and PA, 2008).

Lessons obtained from monitoring and evaluation can be used to transfer knowledge and share lessons learnt nationally and internationally to different stakeholders as well as the public (DN and PA, 2008, Olsen et al., 2005, Mikko, 2009).

Illustrations of Building Back Better: successes and opportunities

Case study: post disaster monitoring and evaluation in Sri Lanka following the 2004 Indian Ocean Tsunami

The 2004 Indian Ocean Tsunami was the most devastating disaster that had impacted Sri Lanka in recent history. There was very little knowledge on disaster management or post disaster recovery in the country at the time. The Post-Tsunami Recovery and Reconstruction Strategy developed by the Sri Lankan Government declared that monitoring mechanisms would be put in place across all sectors subject to recovery programmes, although this was not achieved in practice. The focus was more on speedy reconstruction to re-house the displaced, and less on quality, suitability and sustainability of recovery solutions.

Despite the lack of formal monitoring and evaluation mechanisms, the tsunami experience introduced many new lessons for the improvement of disaster management practices in Sri Lanka. The importance of trying to incorporate disaster risk reduction into structures for the long term, the safety of sites used for construction, and conducting research and model tests to identify high-risk areas have all resulted from lessons learnt following the tsunami.

Lessons from the tsunami experience have led to the implementation of many workshops and conferences to educate the different governmental and non-governmental stakeholders involved.

Case study: monitoring and evaluation following the 2009 Victorian Bushfires

The Victorian Bushfires recovery portrayed several good examples of monitoring mechanisms put in place to improve the recovery effort and measure progress. The Building Commission conducted regular SWOT (Strengths, Weaknesses, Opportunities and Threats) analyses to assess the rebuilding work undertaken. Regular rebuild progress maps were generated in the largest affected town, Marysville, and distributed among key stakeholders to follow the different stages of construction in housing, businesses and infrastructure. The recovery authorities the VBRRA and Force Response Unit conducted surveys to identify progress and to collect information such as the status of the people who had lost their homes and the government's response. The biggest issues were recording of data and difficulties in access and willingness of locals to participate in the surveys.

The supervision of residential rebuilding also raised an issue due to the pressures of rapid rebuilding and shortage of personnel. Insufficient inspections could have resulted in inadequate quality in the rebuild. This highlighted the need to consider resourcing for building supervision during recovery planning.

No notable long-term maintenance or monitoring were in place after the bushfires to manage or monitor hazards. For example, the bushfire risk ratings are dependent on vegetation growth, and it was deemed the responsibility of the homeowner to ensure vegetation was cleared to maintain the hazard risk level. However, there were no inspections or interventions in place to assess this over time.

The Victorian Bushfires Royal Commission was established as a monitoring body to ensure that lessons from the experience were defined and learnt, and several studies were conducted following the bushfires to evaluate the recovery process and draw conclusions. There were previously no proper systems in place to feed lessons learnt back into the system or develop future practices. However, as a result of the bushfires experience, a permanent institution for disaster management, Emergency Management Victoria, was established to retain the lessons and apply them.

Monitoring and Evaluation for Building Back Better

Post disaster monitoring and evaluation has not been the major focus of post disaster reconstruction and recovery efforts. However, BBB advocates for monitoring and evaluation mechanisms to be planned for and put in place in recovery programmes from the outset, to ensure effective and efficient outcomes. Monitoring and evaluation for BBB can be used to ensure compliance as well as to improve ongoing and future recovery activities.

Mechanisms to monitor the quality and compliance of ongoing recovery efforts need to be in place. Monitoring and evaluation includes the collection of comprehensive data about recovery to track progress of recovery outputs. Involving the community more actively in recovery by holding regular meetings or distributing regular newsletters and pamphlets may help the community participate in and lead data collection exercises. Involvement of local councils can also be a good way to obtain local data. Having a good database prepared in the pre-disaster period will assist in recording, analysing and sharing information during post disaster recovery efforts. Post disaster monitoring and evaluation should not be limited to the reconstruction period. Monitoring needs to be put in place in the long term to check on long-term recovery and risk levels on a regular basis.

Monitoring and evaluation can also be used effectively to identify problems with ongoing post disaster recovery practices. Monitoring and evaluation systems provide a good opportunity to evaluate shortcomings in recovery practices in order to learn lessons and improve recovery and disaster management practices for the future. Greater collaboration and open communication between all tiers of government and different stakeholders will enable recovery lessons to reach all parties. The lessons learnt should be factored into revising policies and procedures for future disaster management practices as well as to train stakeholders and the community. It is important to have trained professionals specialising in disaster management and disaster recovery activities who are available to participate in post disaster environments. A database of disaster management/recovery experts can be maintained by the government. If resources allow, having a permanent disaster management section in each council is also recommended.

Greater awareness about disasters and disaster management among communities will facilitate the implementation of disaster management practices. Communities can be kept informed through community meetings, newsletters, brochures, pamphlets, and via radio, television and social media. Information centres can be established to further support the community. Incorporating knowledge about disaster management practices into the education system is also a good way of raising awareness from early ages. The community should be involved in local-level disaster risk reduction and disaster management practices.

BBB Indicators for Monitoring and Evaluation

Indicators or best practices for Building Back Better for post disaster Monitoring and Evaluation were developed based on case study research findings and international examples. Post disaster Monitoring and Evaluation can be used for Compliance with BBB Principles and for Improvement of disaster management and recovery activities. The BBB Indicators for Monitoring and Evaluation are listed in Table 10.1. The Indicators serve as a practical guide to direct stakeholders involved in post disaster activities to understand the elements that need to be considered when planning and implementing reconstruction and recovery programmes in order to Build Back Better.

Table 10.1 Build Back Better Indicators for Monitoring and Evaluation

Compliance	Put in place mechanisms to monitor the quality and compliance of ongoing recovery in line with BBB
	Measure recovery outputs to track progress by collecting comprehensive data about recovery
	Create plans and monitoring and evaluation mechanisms for long-term hazard risk monitoring
Improvement	Identify problems with current post disaster recovery practices through monitoring and evaluation mechanisms
	Incorporate lessons learnt into revising policies and procedures for future disaster management practices
	Train stakeholders on disaster management practices from lessons learnt
	Implement disaster management educational campaigns for the public

References

Business Civic Leadership Center (2012). *What a Successful Disaster Recovery Looks Like*. Available: https://www.uschamberfoundation.org/sites/default/files/publication/ccc/WhatDoesaSuccessfulRecoveryLookLike.pdf [Accessed 9 August 2018].

CERA (2012). *Recovery Strategy for Greater Christchurch*. Christchurch, New Zealand: Canterbury Earthquake Recovery Authority.

Chang, S. E. (2010). Urban Disaster Recovery: A Measurement Framework and Its Application to the 1995 Kobe Earthquake. *Disasters*, 34, 303–327.

Clinton, W. J. (2006). *Lessons Learned from Tsunami Recovery: Key Propositions for Building Back Better*. New York: Office of the UN Secretary-General's Special Envoy for Tsunami Recovery.

Colten, C. E., Kates, R. W. & Laska, S. B. (2008). Three Years after Katrina: Lessons for Community Resilience. *Environment*, 50, 36–47.

Disaster Relief Monitoring Unit of the Human Rights Commission of Sri Lanka (2006). Building Back Better: Way Forward. *National Workshop on Guiding Principles, 2006*. Colombo, Sri Lanka: Practical Action – South Asia Programme. DN & PA (2008). *Disaster and Development in South Asia: Connects and Disconnects. South Asia Disaster Report*. Colombo, Sri Lanka: Duryog Nivaran and Practical Action.

FEMA (2000). *Rebuilding for a More Sustainable Future: An Operational Framework*. Washington, DC: Federal Emergency Management Agency.

GoSL (2005). *Post-tsunami Recovery and Reconstruction Strategy*. Colombo, Sri Lanka: Government of Sri Lanka.

Grewal, M. K. (2006). Sri Lanka – a Case Study. In: *Approaches to Equity in Post-tsunami Assistance*. Colombo, Sri Lanka: Office of the UN Secretary General's Special Envoy for Tsunami Recovery.

Halvorson, S. J. & Hamilton, J. P. (2010). In the Aftermath of the Qa'yamat: The Kashmir Earthquake Disaster in Northern Pakistan. *Disasters*, 34, 184–204.

Kaklauskas, A., Amaratunga, D. & Haigh, R. (2009). Knowledge Model for Post-disaster Management. *International Journal of Strategic Property Management*, 13, 117–128.

Lloyd-Jones, T. (2006). *Mind the Gap! Post-disaster Reconstruction and the Transition from Humanitarian Relief*. London: Royal Institution of Chartered Surveyors.

Matanle, P. (2011). The Great East Japan Earthquake, Tsunami, and Nuclear Meltdown: Towards the (Re)construction of a Safe, Sustainable, and Compassionate Society in Japan's Shrinking Regions. *Local Environment*, 16, 823–847.

Meigh, D. (2009). Aceh Emergency Support for Irrigation – Building Back Better. *Proceedings of ICE Civil Engineering*, 162, 171–179.

Mikko, K. (2009). Managing for Innovation in Large and Complex Recovery Programmes: Tsunami Lessons from Sri Lanka. *International Journal of Project Management*, 27, 123–130.

Monday, J. L. (2002). Building Back Better: Creating a Sustainable Community after Disaster. *Natural Hazards Informer*, 3. Available: www.colorado.edu/hazards/publications/informer/infrmr3/informer3b.htm [Accessed 9 August 2018].

Norio, O., Ye, T., Kajitani, Y., Shi, P. & Tatano, H. (2011). The 2011 Eastern Japan Great Earthquake Disaster: Overview and Comments. *International Journal of Disaster Risk Science*, 2, 34–42.

Olsen, S. B., Matuszeski, W., Padma, T. V. & Wickremeratne, H. J. M. (2005). Rebuilding after the Tsunami: Getting It Right. *AMBIO: A Journal of the Human Environment*, 34, 611–614.

Omidvar, B., Zafari, H. & Derakhshan, S. (2010). Reconstruction Management Policies in Residential and Commercial Sectors after the 2003 Bam Earthquake in Iran. *Natural Hazards*, 54, 289–306.

Quarantelli, E. L. (1999). *The Disaster Recovery Process: What We Know and Do Not Know from Research*. Newark, DE: University of Delaware Disaster Research Center.

Red Cross (2010). *World Disasters Report 2010 – Focus on Urban Risk*. Geneva, Switzerland: International Federation of Red Cross and Red Crescent Societies.

Tas, M. (2010). Study on Permanent Housing Production after 1999 Earthquake in Kocaeli (Turkey). *Disaster Prevention and Management*, 19, 6–19.

Wiles, P., Selvester, K. & Fidalgo, L. (2005). *Learning Lessons from Disaster Recovery: The Case of Mozambique*. Washington, DC: World Bank.

11 Building Back Better
From theory to practice

Introduction

The *Sendai Framework* presented a timely reminder for the future of Building Back Better for the international community. In enacting *Sendai*, there is a need to develop new ways of approaching the recovery environment through a focus on rebuilding with resilience and Building Back Better. *Sendai*'s focus on enhancing disaster preparedness for effective response and Building Back Better in recovery, rehabilitation and reconstruction has been at the heart of this book. Conceptualising Build Back Better in a practical framework with actionable indicators will allow international communities to navigate the complexities of recovery and reconstruction.

This chapter explores how the BBB Framework and Indicators developed through research can be translated effectively into practice. First, a summary of BBB theory is presented, followed by the application of this theory in practice. Feedback from projects where the BBB Framework has been adopted is discussed. Sector-specific use of the tool and its use in different contexts such as natural and man-made disasters and developed and developing countries is presented. The chapter highlights the adaptability of the Framework and its use as a non-prescriptive tool with wide applications.

The theory behind Building Back Better

With increasing exposure and vulnerability to disasters faced by communities, the importance of improving the efficiency and effectiveness of post disaster reconstruction and recovery and using this period to improve the resilience of communities has been well-established. The phrase "Build Back Better" has been used to conceptualise this idea. Many researchers and practitioners have generated research papers and reports bearing suggestions to incorporate BBB practices into post disaster reconstruction and recovery programmes. However, successful implementation of BBB concepts during reconstruction and recovery has been sparse.

The meaning of "Build Back Better" and the various elements required to BBB have been described in many different ways by a number of authors, such as Clinton (2006), Monday (2002) and Kennedy et al. (2008). Different disaster experiences have also contributed towards a range of suggestions to improve recovery efforts in order to introduce future resilience to disaster-affected communities. However, the variety of information found in different sources has been incoherent and incomplete, posing a challenge for true comprehension of BBB and what it entails in a simple and comprehensive manner.

International research and literature uses the term "Build Back Better" to represent a holistic approach towards post disaster reconstruction and recovery, where a community's physical, social and economic conditions are all considered and improved simultaneously to induce a greater level of resilience. Chapter 2 presented the combined findings from analysing successful post disaster reconstruction and recovery practices and BBB theory to identify three key categories required to Build Back Better: (1) Disaster Risk Reduction, (2) Community Recovery and (3) Effective Implementation.

Disaster Risk Reduction was defined as measures put in place to improve the resilience of a community's built environment. Disaster Risk Reduction for BBB was described using three BBB Principles: Structural Resilience, Multi-hazard-based Land-Use Planning, and Early Warning and DRR Education. The second BBB category, Community Recovery, was defined as means of support provided for psychological, social and economic aspects of disaster-affected communities. Community Recovery was described through two BBB Principles: Psychological and Social Recovery and Economic Recovery, which looks at economic rejuvenation. The third category, Effective Implementation, was generated to demonstrate how Disaster Risk Reduction and Community Recovery practices can be put in place in an efficient and effective way. Effective Implementation was described in terms of Institutional Mechanism, Legislation and Regulation, and Monitoring and Evaluation.

Each of the BBB categories and BBB Principles developed in Chapter 2 were detailed and discussed along with case studies illustrating BBB successes and shortcomings in practice in Chapters 3–10. The biggest issues leading to the ineffectiveness of post disaster reconstruction and the failure to achieve BBB-based recovery outcomes were lack of consideration of the wider implications of recovery decisions and lack of practicability in BBB suggestions, which made implementation unfeasible. Common flaws included focusing only on the immediately encountered risk during rebuilds, leading to exacerbated vulnerability to other unanticipated hazards; focusing only on risk reduction, leading to neglected social and economic recovery, poor understanding of the importance of community involvement in recovery; and lack of clarity and the ad hoc nature of recovery efforts, leading to confusion and inefficiency.

The shortcomings, experiences and lessons learnt from the literature and research have provided valuable insights into the issues surrounding BBB and solutions which serve as best-practice recommendations to implement the BBB Principles successfully. These recommendations have been demonstrated as BBB Indicators under each BBB Principle.

Chapter 3 discussed Structural Resilience for BBB and presented BBB Indicators under Building Codes, Cost and Time, and Quality, which were identified as important areas for consideration in implementing structural resilience initiatives during post disaster reconstruction. Despite improving structural resilience seeming straightforward during the rebuild, specific considerations in revising or introducing building codes and building regulations and their corresponding implications on cost, time and quality have major consequences for recovery progress. The BBB Indicators under Structural Resilience highlight the considerations needed to achieve successful BBB outcomes.

Chapter 4 examined Multi-hazard-based Land-use Planning for BBB, with BBB Indicators grouped under Risk-based Zoning and Resettlement. Land-use planning and structural resilience go hand in hand for disaster risk reduction, and need to be considered conjointly to deliver resilient and practical solutions to the community. Resettlement is very disruptive to communities, and can increase their social and economic vulnerabilities despite its aim being to move communities away from disaster risks. Therefore, resettlement needs to be a last resort option, and planned and implemented cautiously and delicately. The BBB Indicators under Resettlement provide guidance on best practices based on international experiences.

The third BBB Principle under Disaster Risk Reduction is Early Warning and DRR Education, and Chapter 5 presented BBB Indicators for these. The BBB Framework identifies the adoption of early warning mechanisms and providing DRR education to the community and stakeholders as equally important DRR measures to build resilience and preparedness within the community.

Chapter 6 addressed an important aspect of post disaster recovery and BBB, Psychological and Social Recovery. BBB Indicators under this BBB Principle recommended putting in place initiatives to support and involve the community as a core aspect of BBB. A community cannot recover without its people, therefore post disaster recovery efforts need to arrange appropriate forms of community support to assist disaster victims to heal from their trauma, re-connect with the community, and make decisions on moving forward. Empowering the community through community involvement in recovery has proven to accelerate psychological and social recovery, therefore recovery efforts must also identify ways in which the community can be involved in the recovery process to some degree.

Chapter 7 presented Indicators for BBB-based Economic Recovery. Similarly to psychological and social recovery, economic recovery is often not a central component in most recovery plans and programmes, which focus mainly on rebuilding. However, economic recovery of a community through the establishment and re-establishment of local businesses and livelihoods is imperative for overall recovery. BBB advocates for economic recovery to be a core component of recovery through the creation of a tailor-made Economic Recovery Strategy, and support through allocated funding for business recovery, assistance with decision-making, training, business support, and business promotion. The BBB Indicators under Economic Recovery provide recommendations on how to incorporate these in recovery planning.

Effective Implementation of the post disaster reconstruction and recovery programme determines the timely, smooth implementation of recovery initiatives and the success of BBB. Chapter 8 introduced the selection of an appropriate

Institutional Mechanism to suit the local community's context as an essential factor for recovery success. The BBB Indicators guide the selection of an appropriate institutional mechanism to suit the community and effective management of recovery stakeholders by fostering partnerships, incorporating grass-roots-level involvement, and quality assurance and training of stakeholders for recovery as well as future disaster management.

Chapter 9 explored the use of Legislation and Regulation to assist BBB. The BBB Indicators illustrate how legislation and regulation needs to be used for compliance of reconstruction and recovery guidelines and initiatives to ensure resilient outcomes, and for facilitation of recovery activities through simplifying and fast-tracking time-consuming procedures.

The final BBB Principle, Monitoring and Evaluation, was presented in Chapter 10. BBB advocates for using effective monitoring and evaluation mechanisms as an indispensable feature in post disaster recovery to ensure compliance with BBB Principles and initiatives, and for improvement of recovery and future disaster management practices through ongoing identification and resolution of problems emerging in the recovery effort. The BBB Indicators under each BBB Principle in Chapters 3–10 advise how previously encountered shortcomings in post disaster reconstruction and recovery can be overcome.

The importance of a holistic approach towards reconstruction recovery in order to Build Back Better was illustrated through the case studies in each chapter. There is a strong link between the recovery of the physical, social and economic environments of communities, which then impact overall recovery. Reconstruction of homes and important community buildings such as schools are necessary for the local community to recover psychologically and feel the drive to move forward and return to their livelihoods and daily routines. Business-owners are only interested in re-establishing themselves in disaster-affected areas based on the extent of rebuilding taking place. At the same time, economic recovery and business re-establishment need to be visible in order to influence affected communities to start rebuilding rather than relocate to different locations. Psychological and social support is required to enable the community to make decisions about rebuilding and return to former livelihoods and daily routines and open or re-open businesses. Absence of any of these elements can affect the other elements, leading to a negative impact on overall recovery.

This book stresses the importance of supporting Disaster Risk Reduction, Community Recovery and Effective Implementation initiatives collectively and synchronically in order to Build Back Better.

Building Back Better in practice

The BBB Framework developed by Mannakkara and Wilkinson provides a holistic view of all the components representing the BBB concept. The Framework is a perfectly aligned tool to achieve the directive set out in the *Sendai Framework* to adopt and implement BBB-based planning for post disaster reconstruction. As a practitioner in either government, non-governmental or private institutions, the BBB Framework and the BBB Indicators developed under each BBB Principle serve as a guide to design post disaster recovery programmes and reconstruction and recovery plans. The Framework, Principles and Indicators bring awareness to critical aspects that need to be considered during reconstruction and recovery in order to Build Back Better and improve a community's resilience.

The BBB Framework and BBB Indicators have been in development for eight years since 2010. The Framework and Indicators were verified by international experts and reconstruction and recovery stakeholders in 2013, and have since continued to be tested and applied in a range of post disaster recovery projects. The Mannakkara and Wilkinson BBB Framework was used to conduct a multi-sector analysis of BBB, testing how BBB Principles and Indicators can apply to and guide resilience and post disaster recovery in a community's social, production and infrastructure sectors through an international study with the World Bank's Global Facility for Disaster Reduction and Recovery. The social sector included housing, education and health, the production sector included employment and livelihoods, and the infrastructure sector included water and sanitation, transport, energy and community infrastructure. BBB in cross-cutting sectors such as environment, gender and governance was also examined. The BBB Framework and the BBB Principles were found to be applicable throughout all these sectors, and served as a reliable umbrella to encompass the core BBB concepts when planning for resilience and recovery. The BBB Indicators were easily modified for each sector, which involved omitting certain Indicators or adding relevant ones. This exercise highlighted the universality of the BBB Framework and its BBB Principles, and illustrated how the BBB Indicators are a non-prescriptive depictive set of best-practice recommendations which can be tailored to different sectors and situations as required.

The BBB Framework and BBB Indicators have been used to assess all aspects of community recovery following major natural disaster events such as the 2004 Indian Ocean Tsunami, the 2009 Victorian Bushfires, the 2010/2011 Canterbury Earthquakes, the 2015 Tropical Cyclone Pam and the 2016 Tropical Cyclone Winston. The BBB Framework, BBB Principles and BBB Indicators were universally applicable in assessing the short- and long-term recovery experiences of both developed and developing countries that experienced these disasters. Regardless of the type of disaster or the country, reducing future disaster risks remains the primary driver in planning a reconstruction and recovery effort. In the BBB Framework, the fundamental ways of reducing risk were improving structural resilience through revised building regulations, hazard-based land-use planning decisions around relocating communities, or a combination of both. Whether through established building codes or less formal building regulations and guidelines, improving structural resilience in both developed and developing countries posed the same challenges around cost, time and quality. Improving structural resilience inevitably increased cost and time in rebuilding, therefore identifying funding sources or alternative risk reduction measures through relocation to lower risk lands needed to be explored in all cases. Quality assurance in rebuilding environments remained difficult in all countries, and common solutions such as attracting skilled builders through the use of incentives and involving local authorities and NGOs in training locals and conducting inspections needed to be applied. Measures for psychological and social recovery and economic recovery also presented common challenges regardless of country or disaster type. The studies represented the universality of post disaster challenges. The comprehensiveness of the BBB Framework and BBB Principles allowed these contrasting disaster contexts to be simplified to identify basic solutions to generate resilient outcomes.

The BBB Framework and Indicators were also used to examine the successfulness of sector-specific post disaster recovery, including studies of tourism business resilience and recovery in the Cook Islands, business recovery in different Christchurch suburbs, and the horizontal infrastructure sector rebuild in Christchurch following the Canterbury Earthquakes. Rebuilding and recovery of the business sector requires Disaster Risk Reduction through improving the structural resilience of physical assets and infrastructure, land-use decisions and early warning and DRR education mechanisms to warn and educate the business community, Community Recovery through assisting the local community and businesses and economic recovery initiatives to support business recovery, and Effective Implementation through understanding the institutional mechanisms and legislation and regulation relevant to business recovery, and Monitoring and Evaluation to assess the progress of business recovery. Infrastructure sector recovery needs to consider all these aspects as well. The recovery of infrastructure impacts directly upon people in relation to restoring access to water, roads, power and telecommunications, and therefore needs to consider social recovery and economic recovery aspects.

A BBB roadmap was developed under the BBB Framework for the recovery and future resilience of agricultural businesses affected by the 2014 conflicts in Gaza, Palestine, commissioned by two local and international NGOs working in the country. The project targeted greenhouse, poultry and livestock and dairy farmers, and fishermen. Despite the man-made disaster context and the unique environment present in Gaza, the BBB Framework and its BBB Principles remained applicable and served as a foundation to modify and develop BBB Indicators to direct the affected agribusinesses towards successful recovery and future resilience.

It is therefore evident that the BBB Framework provides a straightforward guide to identify the key areas that require attention during post disaster reconstruction and recovery in order to Build Back Better and create a resilient post disaster community. Stakeholders involved in planning recovery projects can use the BBB Framework as a guideline to ensure all aspects contributing to BBB are considered and included in recovery plans. The generalised universal BBB Indicators created using first-hand findings from case studies and the international literature provide practical and effective suggestions to implement the BBB categories and Principles introduced in the BBB Framework. Figure 11.1 showcases a proposed sequence of activities in a post disaster context requiring the creation of a new recovery authority.

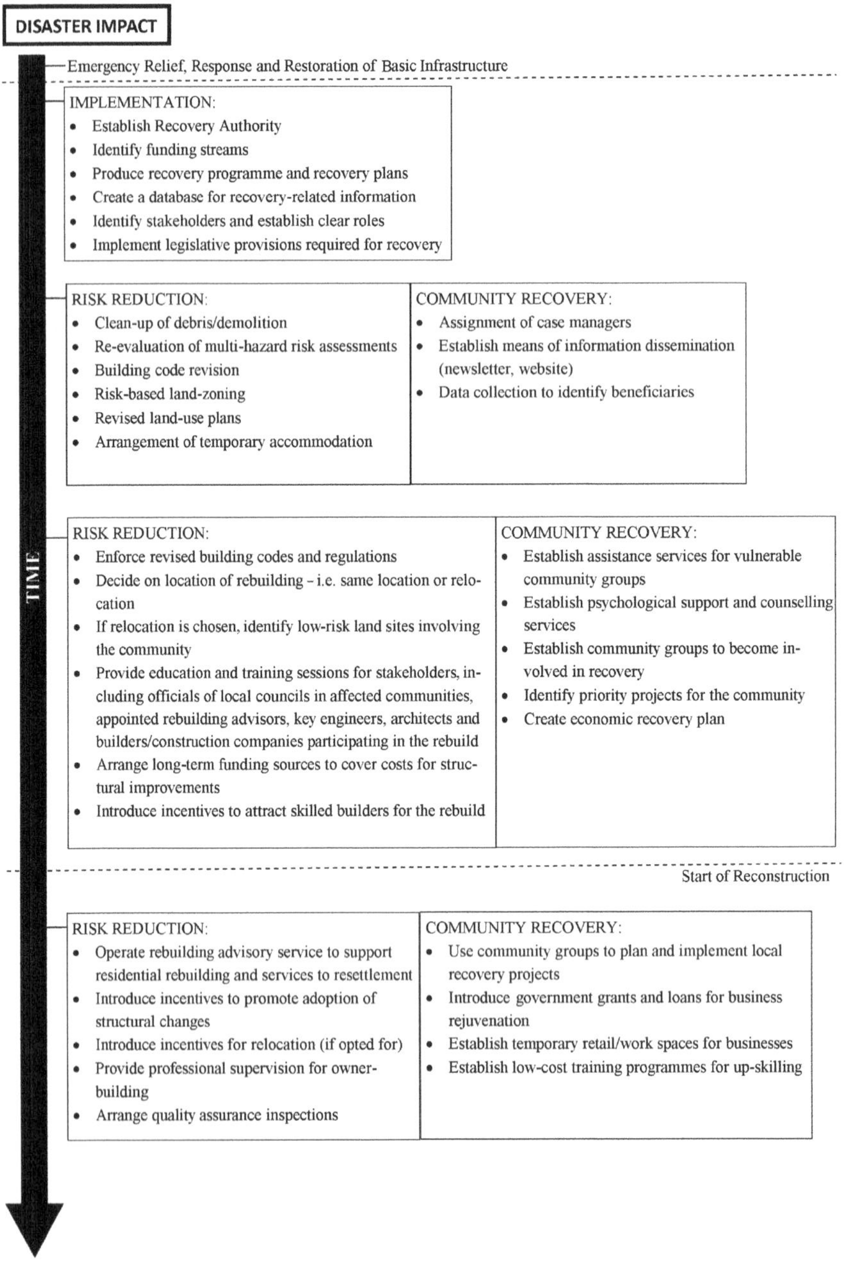

DISASTER IMPACT

— Emergency Relief, Response and Restoration of Basic Infrastructure

IMPLEMENTATION:
- Establish Recovery Authority
- Identify funding streams
- Produce recovery programme and recovery plans
- Create a database for recovery-related information
- Identify stakeholders and establish clear roles
- Implement legislative provisions required for recovery

RISK REDUCTION:
- Clean-up of debris/demolition
- Re-evaluation of multi-hazard risk assessments
- Building code revision
- Risk-based land-zoning
- Revised land-use plans
- Arrangement of temporary accommodation

COMMUNITY RECOVERY:
- Assignment of case managers
- Establish means of information dissemination (newsletter, website)
- Data collection to identify beneficiaries

TIME

RISK REDUCTION:
- Enforce revised building codes and regulations
- Decide on location of rebuilding – i.e. same location or relocation
- If relocation is chosen, identify low-risk land sites involving the community
- Provide education and training sessions for stakeholders, including officials of local councils in affected communities, appointed rebuilding advisors, key engineers, architects and builders/construction companies participating in the rebuild
- Arrange long-term funding sources to cover costs for structural improvements
- Introduce incentives to attract skilled builders for the rebuild

COMMUNITY RECOVERY:
- Establish assistance services for vulnerable community groups
- Establish psychological support and counselling services
- Establish community groups to become involved in recovery
- Identify priority projects for the community
- Create economic recovery plan

Start of Reconstruction

RISK REDUCTION:
- Operate rebuilding advisory service to support residential rebuilding and services to resettlement
- Introduce incentives to promote adoption of structural changes
- Introduce incentives for relocation (if opted for)
- Provide professional supervision for owner-building
- Arrange quality assurance inspections

COMMUNITY RECOVERY:
- Use community groups to plan and implement local recovery projects
- Introduce government grants and loans for business rejuvenation
- Establish temporary retail/work spaces for businesses
- Establish low-cost training programmes for up-skilling

Figure 11.1 Proposed sequence of post disaster recovery activities
Source: Mannakkara and Wilkinson.

The BBB Framework and Indicators can also be used as a guideline to monitor the extent to which BBB practices are implemented during post disaster reconstruction and recovery. Having such a framework as a reference point can assist in improving the success rate of implementing BBB-focused recovery efforts. The BBB Framework also serves as a template to report back on the *Sendai* priority for action on Building Back Better.

Building Back Better depicts the epitome of post disaster reconstruction and recovery success, using activities in the post disaster environment to create resilient communities that are better able to withstand and recover from disaster events. Although BBB has appeared to be in some cases unrealistic to achieve in practice, the Mannakkara and Wilkinson BBB Framework and BBB Principles clarify the core components required to Build Back Better, and the BBB Indicators present non-prescriptive best practices to direct the planning and implementation of recovery, which have made Building Back Better an attainable goal.

References

Clinton, W. J. (2006). *Lessons Learned from Tsunami Recovery: Key Propositions for Building Back Better*. New York: Office of the UN Secretary-General's Special Envoy for Tsunami Recovery.

Kennedy, J., Ashmore, J., Babister, E. & Kelman, I. (2008). The Meaning of "Build Back Better": Evidence from Post-tsunami Aceh and Sri Lanka. *Journal of Contingencies & Crisis Management*, 16, 24–36.

Monday, J. L. (2002). Building Back Better: Creating a Sustainable Community after Disaster. *Natural Hazards Informer*, 3. Available: www.colorado.edu/hazards/publications/informer/infrmr3/informer3b.htm [Accessed 9 August 2018].

Index

Page numbers in *italic* refer to a figure in the text.

accommodation, temporary 88, *126*; and business support 77, *81*, *126*; and legislation and regulation 108, *111*; and psychological and social recovery 66, 67; and structural resilience 37, 40
Aceh, Indonesia 20, 74, 105
ad hoc arrangements 12, 22, 47, 85, 97, 121
"Adopt-a-School" programme 34
agencies, international 94
agricultural businesses 38, 58, 76
Aldrich, D. P. 62
ambiguity 21, 48, 85
American Bar Association 107
Asian Development Bank 93
asset replacement programmes 20, 74
Australia: building codes/regulations 34, 36–7; bushfire-prone areas 14, 36, 108; Canberra Bushfires, Australia 36; community recovery 66, 77, 89; tourism 76, 77; *see also* Victorian Bushfires, Australia
Australian Building Code for Bushfire-prone Areas (AS:3959) 36–7, 108

"Back to Work" programmes 77
Bakir, P. G. 24
Bam Earthquake 22, 24, 114
Bangladesh 92, 93–4; *Bangladesh Climate Change Strategy and Action Plan* 92
Baradan, B. 18, 45
Barpak, Nepal 67
barriers, "hard" and "soft" 104
Basher, R. 54
bathrooms 19, 62
Batteate, C. 16, 20
Bay of Plenty storm, New Zealand 23

Boano, C. 47
Bredenoord, J. and van Lindert, P. 15
Build Back Better 1–9
Build Back Better categories 14–24, 26–7; *see also* community recovery; Disaster Risk Reduction (DRR); implementation, effective
Build Back Better Framework 1, 8, 16–26, 123–7; *see also* Disaster Risk Reduction (DRR); economic recovery; implementation, effective; *Sendai Framework for Disaster Risk Reduction* (SFDRR)
Build Back Better guidelines 12–14; *see also* Clinton, W. J.
Build Back Better Indicators *see* early warning systems; economic recovery; institutional mechanisms; land-use planning, multi-hazard; legislation and regulation; monitoring and evaluation; psychological and social recovery; resilience, structural
Build Back Better Principles 1, 3, 20–2, 24, 25, 121, 123–5; "Build Back Better Guiding Principles", *Post Tsunami Recovery and Reconstruction Strategy* 13
Build Back Better theory 1, 8, 11–27, 120–7; building codes/regulations 15, 17–18, 23, 26, 121–2, 124, *126*; community involvement 14, 15, 19, 26, 121, *126*; community recovery 15, 19–21, 26, 27, 121, 124, 125–6; compliance 18, 23–4, 27, 123; cost constraints 18, 40–1, 121–2, 124; design of buildings 14, 15, 17, 25, 26; Disaster Risk Reduction (DRR) 17–19, 26, 121, 122, 125; early warning systems 17, 26, 121, 122;

Index

economic recovery 15, 20–1, 26, 80–1, 122, 125, *126*; education/training 15, 16, 20, 21, 124, 125, *126*; employment opportunities 15, 16, 23, 124; government, local 13, 18, 21, 22, 125; implementation, effective 21–7, 121, 125, *126*; incentives 125, *126*; Indian Ocean Tsunami disaster (2004) 12, 17–18, 21, 22, 23, 24; infrastructure 14, 26, 124, 125; institutional mechanisms 21, 121, 125, *126*; land-use planning, multi-hazard 15, 50–1, 122, 124, *126*; legislation and regulation 5, 16, 21, 22–3, *126*; lessons learned 16, 23, 24, 121; monitoring and evaluation 16, 21–5, 123, 125; partnerships *13*, 16, 22, 27; psychological and social recovery 15, 19–20, 70–1, 122, 125, *126*; quality standards 40–1, 121–2, 124, *126*; resilience, structural 13, 39–40, 121–2, 124; role allocation 13, 14, 16, 21–2, 27, *126*; skills availability 15, 125, *126*; training programmes 15, 16, 20, 21; vulnerability 12, 15, 24, 121, *126*
building, owner-driven 15, 77, *126*; and institutional mechanisms 89, 90; and psychological and social recovery 62, 66, 70; and structural resilience 35–6, 37
Building Act, New Zealand 106
building codes/regulations 114–15; Australia 34, 36–7, 89, 108; Build Back Better theory 15, 17–18, 23, 26, 121–2, 124, *126*; and enforcement 18, 22–3, 31, 104; Indian Ocean Tsunami disaster (2004) 17–18, 23; and land-use planning 46, 49; and legislation and regulation 21, 23, 103, 104–5, 107, 108, *111*; New Zealand 34, 39, 106; and structural resilience 31–2, 33, 34, 35–7, 39–41, 109, 121–2
Building Commision, New Zealand 115
building consent 20–1, 23, 49, 74
building materials, standard of 12, 108
buildings, public 4, 65, 66, 70, 123
Bureau of Rehabilitation and Reconstruction (BBR), Indonesia 22
Bushfire Attack Levels (BALs), Australia 108
bushfire-prone areas 14, 36, 108
Business Chamber of Commerce, Maryville 77
Business Civic Leadership Centre, "What a Successful Recovery Looks Like" 24
Business Information and Support Service, Australia 77
business recovery 38, 57–8, 110–11, 122, 125; and economic recovery 77, 80, 81
business support 56–7, 69, *111*, 125; businesses, large 56, 57; businesses, small 16, 56, 57, 74; and economic recovery 21, 26, 74, 76–81, 122, 123; and entrepreneurship *13*, 20; and structural resilience 38, 39; and temporary accommodation 77, *81*, *126*
buy-back schemes 45, 49, 108–10

California 24, 106
Canberra Bushfires, Australia 36
Canterbury Earthquake Recovery Authority (CERA), New Zealand 14, 22, 48–9, 78, 79, 96
Canterbury Earthquakes, New Zealand 57–8, 105–6; Christchurch Central City Plan 19, 20–1, 45–6, 74, 79; and economic recovery 20, 74, 78–9, 96, 125; and fast-tracking 20, 74, 96; and infrastructure 78, 79, 96, 125; and institutional mechanisms 78, 87, 96, 96–7; and land-use planning 45–6, 48–9; and monitoring and evaluation 24, 114; SCIRT 49, 78–9, 96–7
Care International and Practical Action 76
case management service 15, 65
cash-for-work programmes 15, 20, 74, 75, 76
Central Business District (CBD), Christchurch 78, 79
Central City Development Unit (CCDU), Christchurch 79
centralisation 85, 86, 87, 97, *100*; *see also* decentralisation
Chamlee-Wright, E and Storr, V. H. 15
Chand Engineering Consultants 34
Chang, K. 15
Chaos Stage 2, 3, 91–2
China 86, 91–2
Christchurch, New Zealand *see* Canterbury Earthquakes, New Zealand
Christchurch Central City Plan 19, 20–1, 45–6, 74, 79
Christchurch City Council 78, 96
Christchurch Earthquake Recovery Authority, *Recovery Strategy for Greater Christchurch* 14
Christchurch-West Melton aquifer system 79
City Care 78, 96

Civil Defence Emergency Management Act, New Zealand 106
climate 5, 23, 40, 54, 92
Clinton, W. J. 1, 31, 53, 120; and economic recovery 15, 21, 74–5; *Key Propositions for Building Back Better* 12–13, 74–5; and monitoring and evaluation 16, 22, 24, 114
coastal communities: "coastal buffer zone" 47–8, 76; and economic recovery 74, 75–6; and land-use planning 18, 45, 47; and structural resilience 30, 31, 35, 39, 91; *see also* Indian Ocean Tsunami disaster (2004)
Coastal Community Resilience Training Workshop, Sri Lanka 91
collaboration 49, 117; and institutional mechanisms 85, 88, 90–2, 93–4, 96, 98–9, *100*
Colten, C. E. 62
Columbia earthquake 19
commemoration 65
Commemorative, Betterment and Developmental Reconstruction Period 2
communication 79, 117; and early warning systems 55, 57, 58; and institutional mechanisms 19, 46, 94, 95, 99, *100*; and land-use planning 19, 46
community, sense of 20, 62–4, 66, 70
community-centred recovery *13*, 14, 15, 66, 79
community consultation groups 20, 70, 110
community involvement 5, 50, 116; Build Back Better theory 14, 15, 19, 26, 121, *126*; and early warning systems 55–6, 57–8; and institutional mechanisms 87, 88–9, 90, 93–4, 98–9, 123; and legislation and regulation 107, 110; and psychological and social recovery 62, 63, 65, 66–70, 88, 122, 123; and structural resilience 33, 34; Victorian Bushfires, Australia 24, 110; *see also* government, local
community recovery 8, 89; Australia 66, 77, 89; Build Back Better theory *13*, 15, 19–21, 26, 27, 121, 124–6; and economic recovery 20–1, 63–4, 77, 81; and psychological and social recovery 19–20, 26, 63–4, 66, 69, 121
community recovery committees (CRCs) 66, 69, 77, 89
Community Recovery Plans 66
community support 50, *100*, 122; and economic recovery 68, 77, 81; and

psychological and social recovery 65–7, 70
competition *13*, 96
compliance 45, 79; Build Back Better theory 18, 23–4, 27, 123; and legislation and regulation 103, 105, 109, *111*; and monitoring and evaluation 114, 117; and structural resilience 31, 34, 36, 37, 123
Comprehensive Rehabilitation and Recovery Plan, Philippines 95
Construction Implementation Unit (CIU), Fiji 34
construction methods 34, 39
Cook Islands 38–9, 56, 68–9, 125
coordination 25, 85, 86, 90, 95
cost constraints 4, 8; Build Back Better theory 18, 40–1, 121–2, 124; and legislation and regulation 104, 108; and structural resilience 31, 32, 33, 36, 37, 40–1
costs 34–5, 39, 45, 57, 87, 96, 109
counselling 19, 21, 66, 68, 69, *126*
cultural acceptability 19, 62
cultural diversity 5
Cyclone Martin 38
Cyclone Sidr 92

Damage and Needs Assessment, Bangladesh 93
data collection 6, *126*; and economic recovery 21, 75, 79–80, *81*; and institutional mechanisms 99, *100*; and monitoring and evaluation 114, 115, 116
Davis, I. 3
deadlocks 77, 80
decentralisation 15, 20, 69; institutional mechanisms 86, 87–8, 95, 97, 98, *100*
decision-making 4, 16, 20, 110, 122; and economic recovery 77, 80–1; and institutional mechanisms 85, 87, 91, 94, *100*; and psychological and social recovery 61, 63, 66, 67, 70
de León, J. C. V. et al. 54
Department of Building and Housing, New Zealand 49
dependency 87
Depression Anxiety Stress Survey (DASS42) 64
design of buildings: Build Back Better theory 14, 15, 17, 25, 26; building design standards 23, 26, 34, 35–6; and early warning systems 56, 57; and infrastructure 78, 79, 96; and

institutional mechanisms 88, 89, 96; and legislation and regulation 107, 108; and seismic resistance 12, 30, 57, 106; and structural resilience 30–7, 39, 41
de Silva, M. W. A. 73, 74
development assistance programmes 6
development planning 12, 45
Diploma in Associate Engineering (DAE), Pakistan 57
Disaster Management Act: Bangladesh 92; Sri Lanka 90–1
Disaster Management Centre (DMC), Sri Lanka 91
disaster management systems 5, 92
Disaster Recovery Framework (DRF), Fiji 33
Disaster Risk Reduction (DRR) 6, 7, 8, 53–60; Build Back Better theory 17–19, 26, 121, 122, 125; *see also* early warning systems; *Sendai Framework for Disaster Risk Reduction* (SFDRR)
Disaster Risk Reduction Education (DRRE) 8, 17, 53–60, 122
DN and PA 19, 46
duplication of effort 16, 22, 25, 104; and institutional mechanisms 86, 88, 98, 100

early warning systems 5, 7, 8, 53–60, 109; Build Back Better theory 17, 26, 121, 122; and communication 55, 57, 58; and community involvement 55–6, 57–8; and design of buildings 56, 57; and education 53–5, 57; evacuation procedures 56, 57–8; and lessons learned 54, 55; preparedness 53–4, 56, 57; and vulnerability 54, 59
Earthquake Reconstruction and Recovery Authority (ERRA), Bangladesh 93–4
earthquake resistance *see* seismic resistance
economic recovery 2, 8, 73–81, 88; Build Back Better theory 15, 20–1, 26, 80–1, 122, 125, *126*; and business recovery 77, 80, 81; and business support 21, 26, 74, 76–81, 122, 123; and Canterbury Earthquakes, New Zealand 74, 78–9, 125; cash-for-work programmes 15, 20, 74, 75, 76; Clinton on 15, 21, 74–5; coastal communities 74, 75–6; and community recovery 20–1, 26, 63–4, 77, 81, 121; and community support 68, 77, 81; and data collection 21, 75, 79–80, 81; and decision-making 77, 80–1; and education 20, 21, 26, 74, 75, 80–1; and fast-tracking 20, 21, 74, 80, 81, 96; and financial assistance 20, 21, 26, 74, 75, 76–7, 80–1; and Indian Ocean Tsunami disaster (2004) 20, 74, 75–6; and inflation 20, 73; and legislation and regulation 23, 106, 109, 110–11, 125; and local government 75, 79–80; and resettlement 73, 74, 75–6; *Sendai Framework for Disaster Risk Reduction (SFDRR) 7*; and skill shortages 80, 81; and Sri Lanka 74, 75–6; and sustainability 15–16, 74, 75; and tourism 74, 75–6, 77, 81; and Victorian Bushfires 76–8, 88
Economic Recovery Team, Victoria 76
education/training 5, 65; Build Back Better theory 15, 16, 20, 21, 124, 125, *126*; and Disaster Risk Reduction 8, 17, 18, 26, 55, 59; and early warning systems 53–4, 57; and economic recovery 20, 21, 26, 74, 75, 80–1; and *Hyogo Framework* 53, 55; and institutional mechanisms 89, 91, 99, *100*; and land-use planning 46, 50, 58; and legislation and regulation 105, 111; and monitoring and evaluation 115, 117; and preparedness 53–4, 56, 57; and public education campaigns 24, 32; schools 33–5, 58; and *Sendai Framework* 54, 55–6
Egbelakin, T. K. 31
Emergency Management Plans, New Zealand 58
emergency management programmes 58, 89, 106, 116
Emergency Management Victoria 89, 116
EMPATHY tool 96
employment opportunities: Build Back Better theory 15, 16, 23, 124; and economic recovery 74, 75, 77, 79, 81; and institutional mechanisms 89, 96
enforcement 18, 22–3, 31, 104
entrepreneurship 13, 20
environment, built *see* building codes/regulations; resilience, structural
environment, natural 2
ethnic groups 62, 67, 76
evacuation procedures 33–4, 39, 56, 57–8
evaluation *see* monitoring and evaluation
Everest, Mount 67
exit strategies 87, 98, *100*
expertise 24, 114, 117; and institutional mechanisms 85, 86, 88, 97, 98, 99

facilitation 23, *111*
families *13*, 63–4, 67

farmers 38, 58, 125
fast-tracking 23, 86; Canterbury Earthquakes 20, 74, 96; and economic recovery 20, 21, 74, 80, *81*, 96; and legislation and regulation 27, 105, 110, *111*, 123
feasibility 95, 121
Federal Emergency Management Agency (FEMA), United States 22, 53
Federal Relief Commission (FRC), Bangladesh 93
Fiji 32–5, 63–5
Fijian Red Cross Society (FRCS) 33, 34, 63–4
Fiji Institute of Engineers (FIE) 34
financial assistance 16, 109, *126*; and economic recovery 20, 21, 26, 74, 75, 76–7, 80–1; and institutional mechanisms 88, 90, 92–6, 99, *100*; and psychological and social recovery 65, 68; and structural resilience 32, 36, 37, 40, 99
Fire Recovery Unit, Victoria 89
fishing industry 20, 74, 75–6, 125
Fletcher Building 96
food sources 64–5
Force Response Unit, Victoria 115
Foreign Aid Transparency Hub (FAITH) 96
FRAMECAD technology 34
Fulton Hogan 78, 96

Gaza, Palestine 38, 58–9, 125
Gazetted Fijian Codes 34
gender awareness 5, 68, 76, 124
Glavovic, B. 46
Global Facility for Disaster Reduction and Recovery, World Bank 124
governance 5, 6, 8, 53, 84–100, 124, *126*; *see also* institutional mechanisms
government, local 68; Build Back Better theory 13, 18, 21, 22, 125; and economic recovery 75, 79–80; and institutional mechanisms 85–6, 91, 95–6; *see also* community involvement
government, national/central 22, 115; and institutional mechanisms 85, 86, 87, 88, 90–6
government organisations 86, 97–8
greenhouses 38
Grewal, M. K. 32
gross domestic product (GDP) 7, 39, 78

Haas, J. E. 2
Haigh, R. 15, 16
Haiti Earthquake 23, 24, 30, 87, 104
Halvorson, S. J. and Hamilton, J. P. 16
Hambantota City Redevelopment Project 47
hazard assessments, multi- 18–19, 54; and land-use planning 45, 47, 48, 49, 50; and structural resilience 31–2, 40, 41
hazard awareness 14, 31, 35, 47, 90, 121
health 7, 63, 64, 124
"Help for Homes" programme 63
Horizontal Infrastructure Governance Group, New Zealand 96
housing 4, 35–7, 64, 68, 90, 106, 124; *see also* accommodation, temporary; building, owner-driven
human factors 55
Hurricane Katrina 20, 55, 63, 86–7, 107–8
Hyogo Framework for Action (HFA) 5–6, 53, 55, 92

ice and snow disaster (2008), China 91
implementation, effective 68; Build Back Better theory 21–7, 121, 125, *126*; and legislation and regulation 26–7, 104, 105, 121; and structural resilience 37, 39; *see also* institutional mechanisms; monitoring and evaluation
incentives 21, 50; Build Back Better theory 125, *126*; and structural resilience 32, 40; tax incentives 14, 32
inconsistency 18, 32, 110
India 20, 115
Indian Ocean Tsunami disaster (2004) 1, 5, 104–5; Build Back Better theory 12, 17–18, 21, 22, 23, 24; building codes/regulations 17–18, 23; and economic recovery 20, 74, 75–6; and institutional mechanisms 86–7, 90–1; and land-use planning 18, 45, 47–8; and monitoring and evaluation 24, 114, 115; and psychological and social recovery 20, 62, 63; and structural resilience 30, 31, 35–6; *see also* coastal communities
Indonesia 20, 74, 105
inexperience 22
inflation 20, 73
information, contradictory 31
information centres 19–20, 65, 117
information sharing 23–4, 79, *126*; and institutional mechanisms 88, 91; and monitoring and evaluation 22, 114,

116, 117; and psychological and social recovery 65, 70
infrastructure 7, 54, 66, 79; Build Back Better theory 14, 20, 26, 124, 125; Canterbury Earthquakes, New Zealand 78, 79, 96, 125; and design of buildings 78, 79, 96; and economic recovery 78, 79; and institutional mechanisms 89, 96; and land-use planning 46, 47; transport infrastructure 23, 105, 124
inspection of buildings 18, 23, 32, 40, 116, 125
institutional mechanisms 84–100; Bangladesh 83–4, 92; Build Back Better theory 21, 121, 125, *126*; and building, owner-driven 89, 90; and Canterbury Earthquakes, New Zealand 87, 96–7; and central government 85, 86, 87, 88, 90–6; and centralisation/decentralisation 85, 86, 87–8, 95, 97, 98, *100*; and collaboration 85, 88, 90–2, 93–4, 96, 98–9, *100*; and communication 19, 46, 94, 95, 99, *100*; and community involvement 87, 88–9, 90, 93–4, 98–9, 123; and coordination 85, 86, 90, 95; and data collection 99, *100*; and decision-making 85, 87, 91, 94, *100*; and design of buildings 88, 89, 96; disaster management systems 5; and duplication of effort 86, 88, 98, *100*; exit strategies 87, 98, *100*; and expertise 85, 86, 88, 97, 98; and government, local 85–6, 91, 95–6; Indian Ocean Tsunami disaster (2004) 86–7, 90–1; and information sharing 88, 91; and infrastructure 89, 96; and leadership 85, 86, 90; and legislation and regulation 90–1, 92, *111*; and lessons learned 87, 97, 99, *100*; and monitoring and evaluation 89, 92, 94, 96, 99, 114, 117; and needs assessment 85, 91, 93, 99; and NGOs 86, 90; and partnerships 88, 90, 95, 98–9, *100*, 123; and preparedness 91–2, 93, 96, 97; and quality 85, 89, 90, 96, 99, *100*; and role allocation 85, 86, 88, 90, 98, *100*; and stakeholders 84, 85–6, 90, 98; and sustainability 87, 92; and time constraints/speed 85, 87, 88, 90, 94, 96; and top-down systems 87, 89; and transparency 85, 94, 95, 96, 99, *100*; Victorian Bushfires 87, 88–90
insurance 32, 37, 39, 57, *81*, 107

International Secretariat for Disaster Reduction-Platform for the Promotion of Early Warning (ISDR-PPEW) 54
International Strategy for Disaster Reduction, UN 54
investment 6, 55

James Lee Witt Associates 19
Japan 19, 24; Earthquake and Tsunami (2011) 20, 74, 87, 114–15
Jha, A. K. 87
Johnson, C. 15

Kashmir Earthquake, Pakistan 12, 30, 57, 92–4
Kennedy, J. 15; with Ashmore, Babister and Kelman 2, 73, 74, 120
Khasalamwa, S. 5, 62
Kijewski-Correa, T. and Taflanidis, A. 2
Kobe Earthquake 30, 63

land-swap schemes 49, *50*, 109–10
land-use planning, multi-hazard 5, 8, 44–51; Build Back Better theory 15, 50–1, 122, 124, *126*; and building codes/regulations 46, 49; Canterbury Earthquakes 45–6, 48–9; and coastal communities 18, 45, 47; and communication 19, 46; Disaster Risk Reduction 17, 25, 26, 121; and education/training 46, 50, 58; and hazard assessments 45, 47, 48, 49, 50; Indian Ocean Tsunami disaster 18, 45, 47–8; and infrastructure 46, 47; and legislation and regulation 50, 103, 106, 107, 108, 109, *111*; and livelihoods 45, 47–8, 50; and permits 45, 50; and resettlement 44–8, 49, *50*, 50, 122; and risk management 45, 46, 49, 50, 122; Samoan Tsunami 18, 45, 46–7; Sri Lanka 45, 47–8; and vulnerability 44, 122
Land Use Recovery Plan, CERA 49
leadership 85, 86, 90, 107
Learning Legacy, New Zealand 96
legislation and regulation 8–9, 103–12, 123; Build Back Better theory 5, 16, 21, 22–3, *126*; and building codes/regulations 21, 23, 103, 104–5, 107, 108, *111*; and community involvement 107, 110; and compliance 103, 105, 109, *111*; and cost constraints 104, 108; and design of buildings 107, 108; and economic recovery 23, 74, 106, 109,

110–11, 125; and education/training 105, 111; and enforcement 18, 22–3, 31, 104; and fast-tracking 27, 105, 110, *111*, 123; Haiti Earthquake 23, 104; and implementation 26–7, 104, 105, 121; and institutional mechanisms 90–1, 92, *111*; and land-use planning, multi-hazard 50, 103, 106, 107, 108, 109, *111*; Pakistan 23, 104; and permits 108, 110; Philippines Disaster Risk Reduction and Management (DRRM) Act 95; and preparedness 106, 107; and quality 108, 110, *111*; and resettlement and temporary accommodation 108, 110, *111*; and risk management 110, *111*; Samoan Tsunami 23, 104; Sri Lanka 23, 104–5; and time constraints 104, 105, 123; Turkey 23, 104; Victorian Bushfires 105, 108–9, 110; and vulnerability 104, 105, 108, 109
lessons learned 47, 75; Build Back Better theory 16, 23, 24, 121; and early warning systems 54, 55; and institutional mechanisms 87, 97, 99, *100*; and monitoring and evaluation 24, 114–15, 116, *117*; Victorian Bushfires 16, 65, 67, 116
Lewis, C. 73
Lewis, J. 2
Listokin, D. and Hattis, D. B. 103
livelihoods 124; and land-use planning 45, 47–8, 50; livelihood recovery programmes 16, 74, 76; livelihood regeneration 15, 20, 25; Sri Lanka 47, 62, 76, 90
Lloyd-Jones, T. 70
loneliness 64
Los Angeles 106
Lovibond, University of New South Wales 64
Lyons, M. 2

Mannakkara, Sandeeka and Wilkinson, Suzanne 8, 19, 21, 63, 73, 123–4
maps 15, 18, 19, 45, 49, 50, 115
marginalised groups 68, 74
Marmara Earthquake, Turkey 19, 62
Marysville, Victoria 66, 76, 77, 115
Marysville Hotel and Conference Centre 77, 78
materials 12, 31, 34, 35, 37, 77, 108
McConnell Dowell 78, 96
Meese III, E. 23
mental health 64, 68

Meyer, M. A. 63
Ministry for Disaster Management, Sri Lanka 91
Ministry of Disaster Management and Relief, Bangladesh 92
Ministry of Education: Fiji 33, 34; New Zealand 58
Mitchell, J. K. 2
Mitigation Plans, Taiwan 19, 45
mobilisation 2, 3–4
Monday, J. L. 21, 24, 75, 120
monitoring and evaluation 9, 54, *111*, 113–17; Build Back Better theory 16, 21–5, 123, 125; and building inspections 18, 23, 32, 40, 116, 125; Canterbury Earthquakes, New Zealand 24, 114; Clinton on 16, 22, 24, 114; and compliance 114, 117; and data collection 114, 115, 116; and education/training 115, *117*; and implementation 27, 121; Indian Ocean Tsunami disaster (2004) 24, 114, 115; and information sharing 22, 114, 116, 117; and institutional mechanisms 89, 92, 94, 96, 99, 114, 117; and lessons learned 24, 114–15, 116, *117*; and quality 114, 115, 116; Sri Lanka 23, 115; Victorian Bushfires 21–2, 24, 115–16
Monitoring and Evaluation Wing (M&E Wing), Bangladesh 94
mortality reduction 7
Motion Consultative Mechanisms 93
multi-hazard approach *see* land-use planning, multi-hazard
Mumbai Floods 12

National Council for Disaster management, Sri Lanka 91
National Development and Reform Commission (NDRC), China 91–2
National Disaster Management Council (NDMC), Bangladesh 92
National Disaster Recovery Framework 53
National Disaster Relief and Recovery Arrangements fund, Australia 77
National DRRM Council (NDRRMC), Philippines 95
National Economic and Development Authority (NEDA), Philippines 95
National Mitigation Strategy, Turkey 14
National Post-tsunami Lessons Learned and Best Practices Workshop, Sri Lanka 23, 105

National Reconstruction Authority (NRA), Nepal 68
needs assessment 20, 21–2, 62; and institutional mechanisms 85, 91, 93, 99
Nepal Earthquake (2015) 67–9
New Normal Stage 2, 4–5
New Orleans *see* Hurricane Katrina
New Zealand 19, 20–1, 23, 58; building codes/regulations 34, 39; infrastructure 78, 79, 96, 125
New Zealand Red Cross (NZRC) 33
New Zealand Transport Agency 78, 96
NGOs 13, 76, 125; and institutional mechanisms 86, 90; and psychological and social recovery 68, 69; Sri Lanka 21, 23, 35, 36, 76, 90; and structural resilience 35, 36
Nokonoko, Fiji 63–4
Northridge Earthquake, Southern California 23, 105, 106–7

Office of the Presidential Assistant for Rehabilitation and Recovery (OPARR), Philippines 95–6
Olsen, S. B. 46
Olshansky, R. B. 70, 87
one-on-one support 19
one-year anniversary 3, 65
overcrowding 64
Ozcevik, O. 70

Pakistan 23, 57, 92–4, 104
Palliyaguru, R. 45
Participatory Flood Risk Communication Support System (Pafrics) 19, 46
partnerships 41, 59, 107–8; Build Back Better theory 13, 16, 22, 27; and institutional mechanisms 88, 90, 95, 98–9, *100*, 123
Pathiraja, M. and Tombesi, P. 74
permits 35, 36, 45, 50, 108, 110
personal agendas 21, 85
Philippines 19, 24, 45, 95–6
Philippines Disaster Risk Reduction and Management (DRRM) Act 95
Philippines Municipal Maps 19, 45
planning *see* early warning systems; land-use planning, multi-hazard; preparedness; resilience, structural
policy-making 16, 74, 86
Post-Disaster Recovery Framework (PDFR), NRA 68
post-traumatic stress disorder 61

Post-Tsunami Recovery and Reconstruction Strategy, Sri Lanka 13, 75–6, 115
Potangaroa, R. 74
poultry 38
poverty 20, 64, 73
preparedness 2, 5, 6, *13*, 40, 114; and early warning systems 53–4, 56, 57; and education/training 53–4, 56, 57; and institutional mechanisms 91–2, 93, 96, 97; and legislation and regulation 106, 107
Principles for Settlement and Shelter 13, 24
Priority Implementation Partnership (PIP) projects 90
Project Implementation Coordination Units (PICUs), Bangladesh 94
prolonged grief disorder 61
psychological and social recovery 8, 61–71, 109; and accommodation, temporary 66, 67; Build Back Better theory 15, 19–20, 70–1, 122, 125, *126*; and community 19–20, 26, 62–70, 88, 121, 122, 123; and counselling 19, 66, 68, 69; and decision-making 61, 63, 66, 67, *70*; and families 63–4, 67; and financial assistance 65, 68; Indian Ocean Tsunami disaster (2004) 62, 63; and information sharing 65, *70*; and NGOs 68, 69; and social cohesion 62, 67–9, *70*; Victorian Bushfires 15, 19, 65–7; and vulnerability 62, 64, 68, 69, 70
public education campaigns 24, 32

quality standards 62; Build Back Better theory 40–1, 121–2, 124, *126*; and institutional mechanisms 85, 89, 90, 96, 99, *100*; and legislation and regulation 108, 110, *111*; and monitoring and evaluation 114, 115, 116; and structural resilience 31, 32, 35, 40, 121–2
Quetta Earthquake 57

Ra High School, Fiji 33
Rakiraki, Fiji 63
Realisation Stage 2, 3
Rebuilding Advisory Service (RAS), Victoria 37, 65, 77, 89
Reconstruction and Development Agency (RADA), Sri Lanka 35, 90
Reconstruction and Recovery Fund (RRF), Yemen 94–5
Reconstruction Supreme Supervisory and Policymaking Association (BRSSPA), Iran 22

"Recovery and Reconstruction Framework", Victoria 14
recovery authorities 22, 27, 84–100, 109, 111; *see also* institutional mechanisms
Recovery Strategy for Greater Christchurch, CERA 14
Red Crescent 13
Red Cross 13, 14, 16, 33, 75; Fijian Red Cross Society (FRCS) 33, 34, 63–4
regulation *see* building codes/regulations; legislation and regulation
Replacement Reconstruction Period 2
resettlement 16, 19, 35, *126*; and economic recovery 73, 74, 75–6; and land-use planning 44–8, 49, 50, *50*, 122; and legislation and regulation 108, 110, *111*; Samoan Tsunami 46–7, 62; Sri Lanka 47–8, 62, 75–6
resilience, definition 5
resilience, structural 8, 30–41; and accommodation, temporary 37, 40; Build Back Better theory 13, 39–40, 121–2, 124; and building, owner-driven 35–6, 37; and building codes/regulations 31–2, 33, 34, 35–7, 39–41, 109, 121–2; and business support 38, 39; and coastal communities 30, 31, 35, 39, 91; and community involvement 33, 34; and compliance 31, 34, 36, 37, 123; and cost constraints 31, 32, 33, 36, 37, 40–1; and design of buildings 30–7, 39, 41; and evacuation procedures 33–4, 39; and financial assistance 32, 36, 37, 40, 99; and hazard assessments 31–2, 40, 41; and implementation 37, 39; and incentives 32, 40; Indian Ocean Tsunami disaster (2004) 30, 31, 35–6; and insurance 32, 37, 39; and materials 31, 34, 35, 37; and new technology 31, 34, 38; and NGOs 35, 36; and permits 35, 36; and quality standards 31, 32, 35, 40, 121–2; and skill shortages 34, 35, 37–8, 40; and speed/time constraints 31–2, 35, 36, 37, 121–2; and Sri Lanka 31, 35–6; and vulnerability 31, 33, 35, 41
Resource Management Act (RMA), New Zealand 106
response capability 55
Restoration Period 2
risk assessment 49, *126*
risk-based zoning 50, 122
risk management 5, 19, 30, 55; and land-use planning 46, 49; and legislation and regulation 110, *111*

rivalry *13*
role allocation: Build Back Better theory 13, 14, 16, 21–2, 27, *126*; institutional mechanisms 85, 86, 88, 90, 98, *100*
Rubin, C. 85
rural areas 18, 31, 57

Samaratunge, R. 85
Samoan Tsunami 18, 30, 62, 74; and land-use planning 18, 45, 46–7; and legislation and regulation 23, 104; and resettlement 46–7, 62
schools 33–5, 58, 64; *see also* education/training
seismic resistance 12, 30, 57, 106
Sendai Framework for Disaster Risk Reduction (SFDRR) 1, 6–7, 11, 92, *117*, 123, 127; and education 54, 55–6
separation 19
settlements, informal and illegal 14; *see also* resettlement
Sixth Five-year Plan of Bangladesh 92
skills availability 65, 114; Build Back Better theory 15, 125, *126*; and economic recovery 77, 80, *81*, 89; skills training programmes 15, 77; and structural resilience 34, 35, 37–8, 40
Small Business Mentoring Service, Australia 77
social bonding 63, 67–8
social cohesion 19, 20, 62, 67–9, *70*
South Asia Disaster Report 14–15
South Pacific Convergence Zone (SPCZ) 38
speed 19, 25, 62, 115; institutional mechanisms 85, 90; and resilience, structural 31–2, 35, 36, 37
Sri Lanka: and economic recovery 74, 75–6; and institutional mechanisms 13, 90–1; and land-use planning 45, 47–8; and legislation and regulation 23, 104–5; livelihoods 47, 62, 75–6, 90; monitoring and evaluation 23, 115; and NGOs 21, 23, 35, 36, 76, 90; resettlement 47–8, 62, 75–6; and resilience, structural 31, 35–6; *see also* Indian Ocean Tsunami disaster (2004)
stakeholders 6, 84, 85–6, 90, 98, *126*
standards, building design *see* building codes/regulations
standard setting 93
State Overall Planning (SOP) for Post-Wenchuan Earthquake Restoration and Reconstruction 91–2

Stronger Christchurch Infrastructure Rebuild Team (SCIRT) 49, 78–9, 96–7
Struggle Stage 2
suitability 87, 115
sustainability 21, 106, 115; and economic recovery 15–16, 74, 75; and institutional mechanisms 87, 92
SWOT(Strengths, Weaknesses, Opportunities and Threats) analysis 115

Tamil Nadu, Sri Lanka 63
Tas, M. 23
Task Force to Rebuild the Nation (TAFREN), Sri Lanka 35, 90
tax incentives 14, 32
Technical Advisory Groups, Bangladesh 94
technologies, new 31, 34, 38, 58
Thiruppugazh, V. 87
time constraints 121–2; and institutional mechanisms 87, 88, 94, 96; and legislation and regulation 104, 105, 123
Tokyo earthquake (1923) 62
top-down systems 55, 87, 89
tourism: Australia 76, 77; Cook Islands 39, 56, 68–9, 125; and economic recovery 74, 75–6, 77, 81
Tourism Chambers, Marysville 77
Tourism Industry Council, Rarotonga 56
town planning 31
tradespeople 32, 34, 37, 62, 89
training *see* education/training
training programmes 40, 91; Build Back Better theory 15, 16, 20, 21; and economic recovery 74, 75, 80
transparency 70, 85, 94, 95, 96, 99, *100*
transport infrastructure 23, 105, 124
Tropical Cyclone Winston, Fiji 32–5, 63–5
Tsunami Evaluation Commission 12
Tsunami Recovery Impact Assessment and Monitoring System (TRIAMS) 24, 114
Turkey 14, 19, 23, 62, 104
Twigg, J. 22, 55
Typhoon Yolanda 95

United Nations 1, *13*; United Nations Disaster Relief Organization 13, 24; United Nations Environment Programme 11

Urban Development Authority, Sri Lanka 47
US Federal Emergency Management Agency 14, 107

vegetation growth 116
Victorian Bushfire Appeal Fund (VBAF) 37, 65, 109
Victorian Bushfires, Australia 36–8, 45; and buy-back schemes 45, 108–9; and community involvement 24, 110; and economic recovery 76–8, 88; and institutional mechanisms 45, 87, 88–90, 109; and legislation and regulation 105, 108–9, 110; and lessons learned 16, 65, 67, 116; and monitoring and evaluation 21–2, 24, 115–16; and psychological and social recovery 15, 19, 65–7; Victorian Bushfire Reconstruction and Recovery Authority (VBRRA) 22, 37, 65, 66, 76–7, 88–90, 115; Victorian Bushfires Royal Commission 14, 16, 21–2, 106, 116; Victorian Government 45, 88, 89, 109
volunteer participation 5, 33, 64, 67
vulnerability 2; Build Back Better theory 12, 15, 24, 121, *126*; and early warning systems 54, 59; and land-use planning 44, 122; and legislation and regulation 104, 105, 108, 109; and psychological and social recovery 62, 64, 68, 69, 70; and structural resilience 31, 33, 35, 41
Vunikavikaloa Arya Primary School, Fiji 33–5

water supplies 38, 47, 76, 78, 79, 124; *see also* infrastructure
Wenchuan Earthquake, China 86, 91–2
"What a Successful Recovery Looks Like", Business Civic Leadership Centre 24
Williamson, S. 2–5, 8
Winchester, P. 16
wind resistance 33–4
World Bank *13*, 93, 124
World Conference on Disaster Reduction 5, 6
World Disasters Report 2010 14
World Health Organisation 94
World Heritage Sites, UNESCO 67

Yemen 94–5

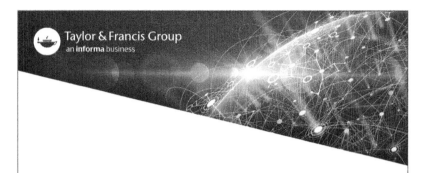